The Evolution of Aging

The Evolution of Aging

How New Theories Will Change the Future of Medicine

Third Edition

THEODORE C. GOLDSMITH

Azinet Press
Box 239 Crownsville, MD 21032
books@azinet.com

ISBN: 978-0-9788709-5-9

Kindle version: ASIN B001E5DXOK
Barnes and Noble Nook version: BN ID: 2940012061270

Web version: http://www.azinet.com/aging/Aging_Book.pdf

Print Versions:
The Evolution of Aging, Third Edition
ISBN: 978-0-9788709-5-9 4/2014
The Evolution of Aging Second Edition, Theodore C. Goldsmith
ISBN: 0-9788709-0-5 10/2006
First edition: ISBN: 0-595-28069-2 6/20/2003 trade paperback

Cover art: Bulent Ince.

Keywords: ageing, antiaging, senescence, regenerative medicine, gerontology, evolution, health, biology, anti-aging medicine, science

April 13, 2014

Summary:

Why do we age? This question has baffled scientists for 150 years and remains unresolved. Scientists disagree over even the general nature of aging. Is aging the result of fundamental limitations that apply to all living things, do we age because our bodies do not try harder not to age, or are organisms designed by nature or "programmed" to age because a limited lifespan conveys some evolutionary advantage? Amazingly, in the 21st century, there is still scientific disagreement regarding this question.

This issue has potentially enormous implications for medicine. If aging is the result of fundamental and unalterable forces of nature, then anti-aging medicine is impossible and anti-aging research is futile and foolish. If aging is dictated by a species-specific design, then research may well reveal means for altering the operation of the aging mechanism and thereby improve the treatment of many age-related diseases and conditions.

This book provides a review of theories of biological aging including underlying evolution and genetics issues and describes recent discoveries and theories that overwhelmingly favor programmed aging and therefore suggest that increased research on aging mechanisms would be highly beneficial to public health.

Acknowledgements

The author wishes to thank the many people who helped with this book or provided helpful comments and suggestions regarding the underlying material including Joshua Mitteldorf (University of Arizona), Vladimir Skulachev (Moscow State University), Jeff Bowles, Aubrey de Grey (University of Cambridge), Vicky Cahan (National Institute on Aging), and Joao Magalhaes (Harvard Medical School).

Thanks also to Frann Goldsmith for her editing assistance and Elaine Bailey (Trident Press International) for editorial comments.

Photo and Illustration Credits

Photos of Darwin, August Weismann, Peter Medawar, and Galapagos tortoise from Wikipedia.

Science Abstract:

In 1882 Weismann formally proposed that programmed aging was an evolved characteristic, an adaptation of organism design, which had an evolutionary purpose. Darwin had previously suggested that aging was an evolved characteristic despite conflicts with Darwinian evolutionary mechanics.

Many current theorists have discarded adaptive theories of aging using one or more of the following arguments: Some theorists consider it "impossible" that aging could be an adaptation because adaptive theory conflicts with their theory of natural selection. Others rejected adaptive theories because no propagation mechanism had been demonstrated that would provide for the evolution of a fitness-adverse trait such as aging. Finally, some doubt that aging has evolutionary utility.

This book, written for a wide audience, presents arguments that refute all of these objections, concludes that programmed aging is the correct theory beyond a reasonable doubt, and provides a summary of modern programmed aging theories as well as an analysis of the potential public health implications of adaptive theories.

Preface

I am often asked: "Why is digital communications theory important to genetics, evolution, and theories of biological aging?" Biology and space communications certainly seem to be entirely unrelated.

However, there is actually an important and very interesting connection. All living organisms have the ability to communicate their design characteristics to their descendants by means of their genetic codes. This inter-generational genetic digital communications system is strikingly similar to modern space communications systems. The genetic code is a serial digital data stream. Nature and NASA face similar issues in areas such as data redundancy, error recovery, synchronization, and data merging, and have developed similar solutions. Both systems have benefits and limitations conferred by the fundamental nature of digital communications.

Evolution is the process whereby genetic codes are modified, initially as the result of "transmission error." Evolution is therefore highly dependent on the details of the genetic communications and data storage scheme including the digital benefits and limitations.

Finally, as the title of this book implies, theories of aging and anti-aging research are very constrained by evolution theory.

The merging of biology and information technology has spawned a number of new fields of science including *genomics*, and *proteomics*, under the general term *bioinformatics*. My main area of scientific interest concerns the consequences of the digital nature of the genetic communications system on evolution theory and thereby aging theory.

Some say: "Darwin was wrong." Others take the position that Darwin was 100 percent correct and that any valid deviation is "impossible." This book presents the case that Darwin was perhaps 99 percent correct. Digital genetics analysis and other evidence suggest that minor adjustments to Darwin's theory are necessary. Although these changes have little effect on our general understanding of evolution, they have a crucial effect on aging theory.

Theodore. C. Goldsmith
tgoldsmith@azinet.com
Annapolis, Maryland

Contents

Table of Figures

1. Introduction

Why do we age?

There are currently three main theories of biological aging:

1) Aging is the result of fundamental limitations such as laws of physics or chemistry that result in deterioration of any living organism. These are the ***fundamental limitation theories*** or "wear and tear" theories.

2) Modern ***non-programmed aging theories*** dating from the 1950s say that aging occurs because our bodies do not better oppose natural deteriorative processes. Tissue damage therefore accumulates causing aging.

3) Modern ***programmed aging theories*** dating from the 1990s say that we age because we possess what amounts to a genetically programmed suicide mechanism that purposely limits our lifespan.

Although the first idea is still popular among the general public, most gerontologists and other medical researchers now believe in one of the last two theories because they provide a much better match to major observational evidence.

Evolutionary mechanics theory or the theory of how the evolution process operates is crucial to aging theories. Everybody learns about Darwin's survival-of-the-fittest evolutionary mechanics concept in introductory biology venues. About 99 percent of all observations ever made concerning characteristics of living organisms fit with Darwin's mechanics concept, which also fits with the fundamental limitation theories. People who have not studied modern evolutionary mechanics theories therefore tend to logically believe in the fundamental limitation theories.

However, despite more than 90 years of effort theorists were unable to explain the huge observed variation in organism lifespans based on fundamental limitations. As examples, mammal lifespans vary over a range of more than 200 to 1 between Argentine desert mouse (< 1 year) and bowhead whale (> 200 years) and fish lifespans vary over a range of more than 1300 to 1 between Eviota sigillata (~ 8 weeks) and Koi (> 200 years). Similar organisms are constructed from similar materials (meat, bone, blood, etc.) and nobody could figure out a plausible reason why fundamental and universal limitations would affect such similar materials so differently as to result in the huge lifespan variations. After nine decades it became clear that modifications to Darwin's mechanics were necessary to solve the aging problem. ***All*** of the modern programmed and non-programmed aging theories are consequently based on modifications to Darwin's survival-of-the-fittest concept to be described in detail and all of these theories provide explanations for the lifespan variations. Eventually other discrepancies were discovered leading to yet more modifications. Where Darwin proposed a single evolutionary mechanics concept there are now multiple concepts. Where the mechanics of evolution used to be seen as simple and elegant they are now seen as complicated and messy.

For many decades, programmed aging theories were considered to be theoretically impossible based on the mechanics of the evolution process and many scientists still favor non-programmed aging theories. In 1960 it was literally true that programmed aging was theoretically impossible because at the time there was no evolutionary mechanics theory support for programmed aging and there was a substantial evolution theory rationale to the effect that a suicidal organism design could not be produced by the evolution process, at least not in mammals. In addition there was relatively little direct evidence of programmed aging especially in mammals.

However, since then this situation has dramatically changed in favor of programmed aging theories as this book will describe in detail. There are now multiple evolutionary mechanics theories that support programmed aging in humans and other mammals. Direct evidence of complex programmed aging mechanisms has been discovered in multiple organisms. In addition, it is now widely understood that the evolution process is much more complicated and less well understood than previously thought. The previously popular idea that we understood the evolution process so perfectly that we could declare programmed aging "impossible" on evolutionary mechanics grounds despite any amount of direct contrary evidence is no longer scientifically supportable.

Which aging theory is correct has potentially huge implications for medicine. Most deaths and the majority of medical expense in developed countries are the result of age-related diseases. As will be described, programmed and non-programmed theories have very different predictions regarding the nature of age-related diseases and therefore the best approaches toward treating and preventing these diseases.

Many people are amazed to find that in the twenty-first century there is still scientific disagreement regarding the fundamental nature of aging. The problem has not been lack of clues. The problem is that the many clues point in different directions resulting in drastically different conclusions. In addition, compared to most areas of scientific inquiry, aging is surrounded by non-science factors that tend to confuse both popular and scientific thought. Understanding the current situation therefore requires an understanding of these confusion factors as well as the historical sequence in which scientists made important discoveries and developed important theories. A major portion of this book is devoted to covering these aspects of aging theory.

Most medical advances have been the result of experimentation. Many major discoveries have been essentially accidental. Aspirin works. We found that out by trial and error. Why it works is of interest, but secondary.

Aging, however, because it is a relatively long-term process, is a difficult subject for experimentation. An experiment to determine if a medication is effective in relieving pain, increasing kidney function, or suppressing a certain infectious organism, could be performed in a matter of days or weeks. An experiment to determine if a medication increases longevity in animals or humans could take years, decades, or even many decades to perform.

As a consequence, scientific theories of aging are primarily the result of logical analyses of the functional, externally observable, characteristics of various organisms. You can readily picture the difficulties associated with this approach. Imagine trying to deduce the existence of, much less the detailed functioning, of the endocrine system, or other largely internal system, merely by observing how animals live and die. It is difficult or impossible to prove such a logical theory without experimentation, which remains difficult.

In any field, theories and assumptions that have remained unchallenged for relatively long periods tend to acquire the appearance of facts. NASA uses an analytic technique called *zero-base analysis* in which previously made assumptions are reevaluated in light of current information. This book is essentially a zero-base examination of scientific and popular beliefs regarding aging.

Another typical engineering process applied in this book is *limit analysis* in which scientists vary the parameters of a system model and assess the results. "In the limit" a parameter can be set to zero or infinity. In the current context, we might explore what would happen to an animal population if the animal's lifespan were longer or shorter or if it did not age at all.

Finally, the development of modern digital communications systems has led to insights into the properties and issues associated with the fundamental nature of digital data. The mechanism of inheritance in all living organisms involves the transmission of digital data in the form of genetic codes. The properties of digital data therefore constrain inheritance, which in turn constrains evolution, which in turn affects aging theory. Accordingly, this book provides a discussion of *digital genetics* or the impact of the digital nature of inheritance on evolution theory.

What Is Aging

Aging refers to the time-sequential deterioration that occurs in most animals including weakness, increased susceptibility to disease and adverse environmental conditions, loss of mobility and agility, and age-related physiological changes. Aging is usually understood to include reductions in reproductive capacity. In this book, it is assumed that aging includes changes in reproductive capacity including behavioral patterns such as reproductive vigor or strength of the urge to mate. Reproductive capacity and aging have similar theory issues as will be discussed later.

Aging and lifespan refer to *internal* limitations on an organism's lifetime as opposed to *external* limitations such as predators, environmental conditions, food supply, habitat, or infectious diseases. The relationship between internal and external limitations is a major consideration in modern aging theories.

A non-aging animal is such a foreign concept that discussing it in English is awkward. Some scientists use the term *senescence* to indicate the deteriorating effects of aging as opposed to the simple passage of time. In this book, *aging* means the deteriorating effects. A non-aging animal does not age but does get older and has increasing "calendar age."

We will be examining two major questions about aging:

1. What is the fundamental nature of aging? Is aging a *feature* of an organism's design that evolved because it confers some benefit, or, is aging not an evolved characteristic but rather a limitation, a *defect* that confers no benefit? Is aging purposely *programmed* by the organism's genetically programmed design? Is aging the *passive non-programmed* result of forces acting upon the organism or the *active* result of a pro-aging design?
2. Are there potentially medically treatable factors that are common to the various manifestations of aging such that a single treatment could delay or ameliorate many different manifestations? Is aging itself a potentially treatable condition or are we limited to treating the various manifestations separately? Is *anti-aging medicine* possible?

Scientific attitudes regarding these questions vary dramatically depending on discipline. Some biologists tend to believe that it is impossible for aging to be an evolved characteristic.

(Biologists have been arguing about this question for more than 150 years.) Medical scientists are more likely to accept either answer for question 1 but consider the question itself to be very academic, "What difference does it make?" As we will see, question 1 has very significant implications for question 2, which does have obvious impact on medicine.

Human Mortality

Developed countries have been keeping fairly careful birth and death records for hundreds of years partly because this data is central to calculations involved in life insurance, pensions, and annuities. The curve shown below shows mortality[1] in the United States in 1999 as a function of age. That is, the curve shows the fraction (equivalent to probability of death) of the people that age that died in 1999. The curve shows that the probability of dying approximately increases *exponentially* with age. The chance of dying during any given year is less than 1 percent until age sixty and rises to 30 percent at age ninety-seven.

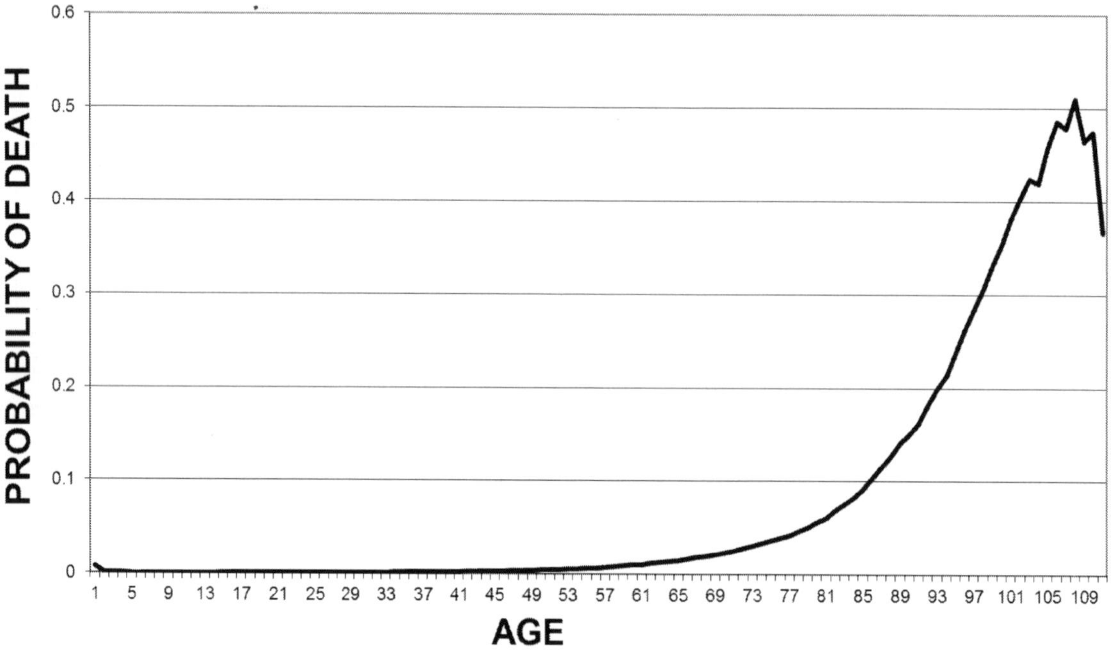

Figure 1 Probability of Death vs Age

The graph below shows the same information displayed on a logarithmic vertical scale to make it easier to see the detail at the lower ages. Data for males (top curve) and females (bottom curve) has been added to the curve for the total population (middle curve). On this figure you can clearly see the infant and early childhood mortality that declines to approximately age 4, and the childhood period where death rates are extremely low (< .02 percent). Then we see a sudden jump to another plateau between ages 18 and 30 where death rates are about .1 percent and where a dramatic difference between males and females is apparent. From age 30 onward, death rates nearly follow the exponential curve, doubling approximately every ten years. A 30 year old has a .1 percent chance of dying while 30; a 101-year-old has a 40 percent chance of dying before reaching 102. Biologists have long been curious as to why death rates level off for very old people.

Some actuaries use a simple exponential equation to approximate the probability of death as a function of age. This *Gompertz Approximation* (developed by Benjamin Gompertz 1779 – 1865) would appear as a straight line on the graph below and is useful since the deviations between the actual curve and a straight line at very young and very old ages are relatively insignificant in an actuarial context.

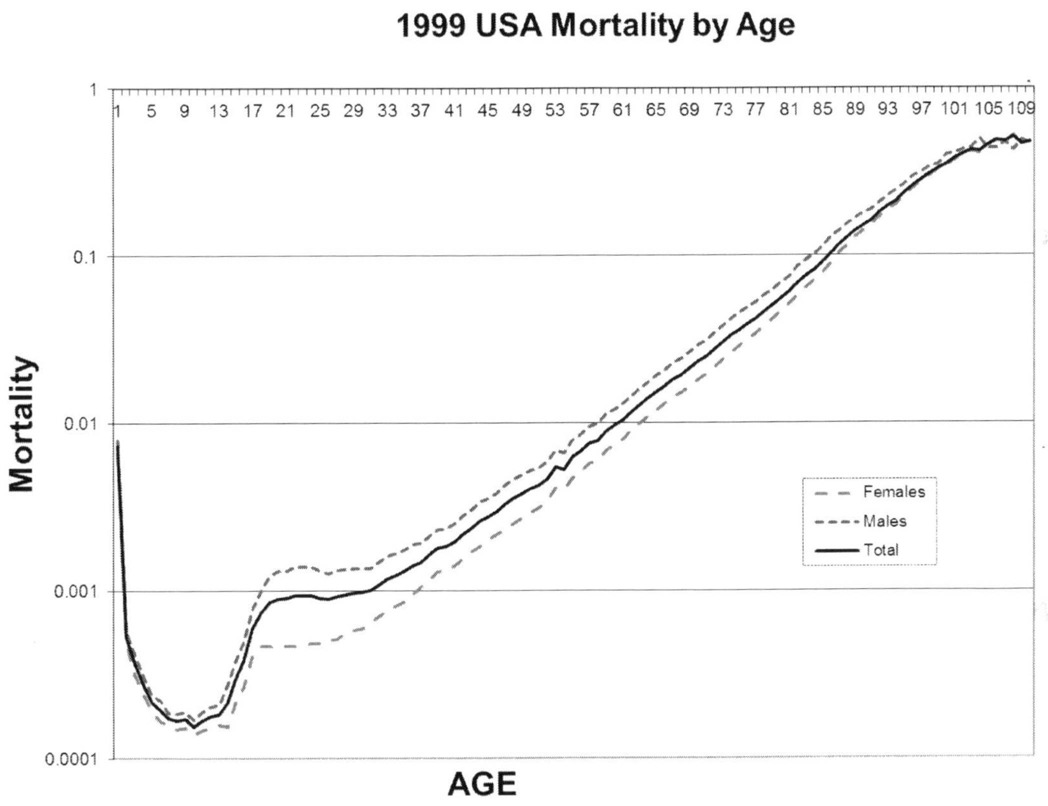

Figure 2 1999 USA Mortality by Age

The next graph below again shows the same data as a survivor's curve based on the ages of people who died in 1999 and showing the fraction of the people left alive as a function of age. This curve shows that 90 percent of the people who died in 1999 were more than fifty-eight years old. Of those that died, 50 percent were more than seventy-eight years old.

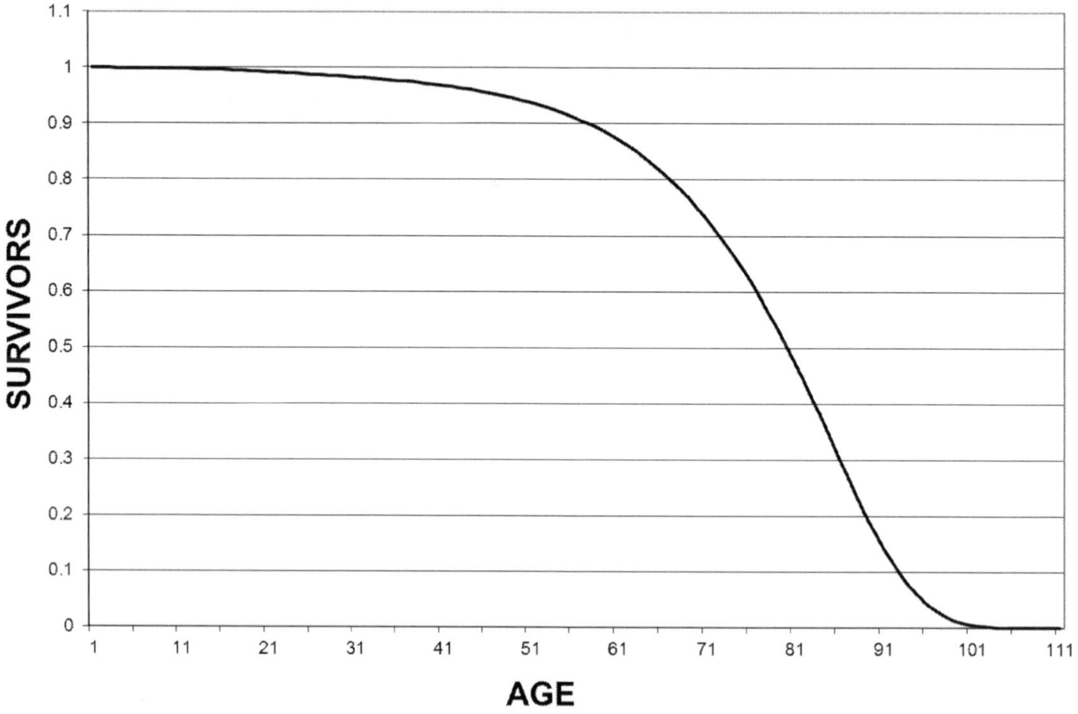

Figure 3 1999 USA Survivors vs. Age

The graph below compares U.S. deaths per 100,000 based on information from the Human Mortality Database for deaths in 1933, 1999, and 2010. This graph reflects total deaths as opposed to death rate and shows that the largest number of people died at 88 years of age in 2010 and about 1 percent at age 100.

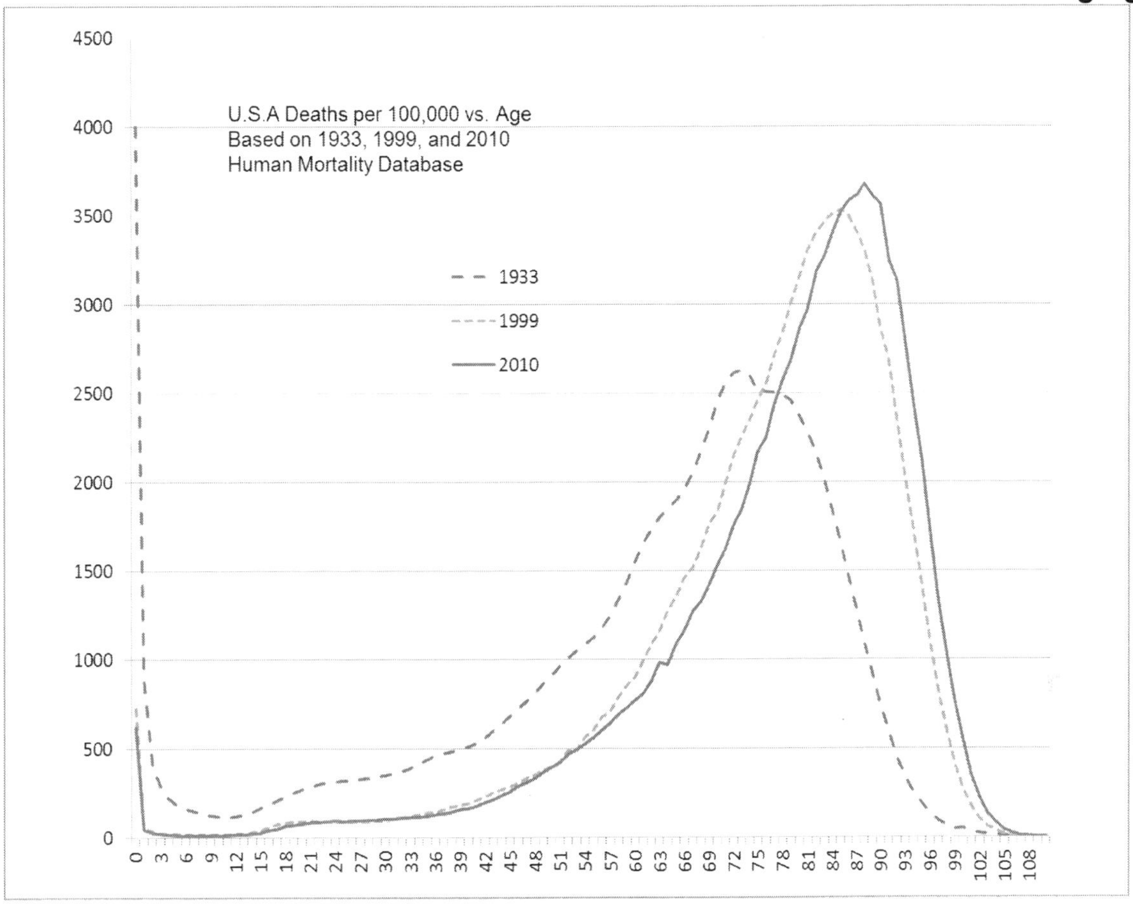

Figure 4 USA Deaths vs. Age for 1933, 1999, and 2010

We can see that between 1933 and 1999 there was a very significant decrease in infant mortality as well as major decreases in childhood and early adult mortality. Between 1999 and 2010 there was relatively little change in mortality for people less than 53 years old. However, for people between 53 and 85 there was a rather amazing delay in mortality given the short time interval between 1999 and 2010.

It is not clear what caused the noticeable discontinuity at age 63 in the 2010 data that corresponds to birth in 1947. Perhaps this is an artifact of World War II.

The graph below shows mortality data for the U.S. and Japan for 2010 and 2009 respectively.

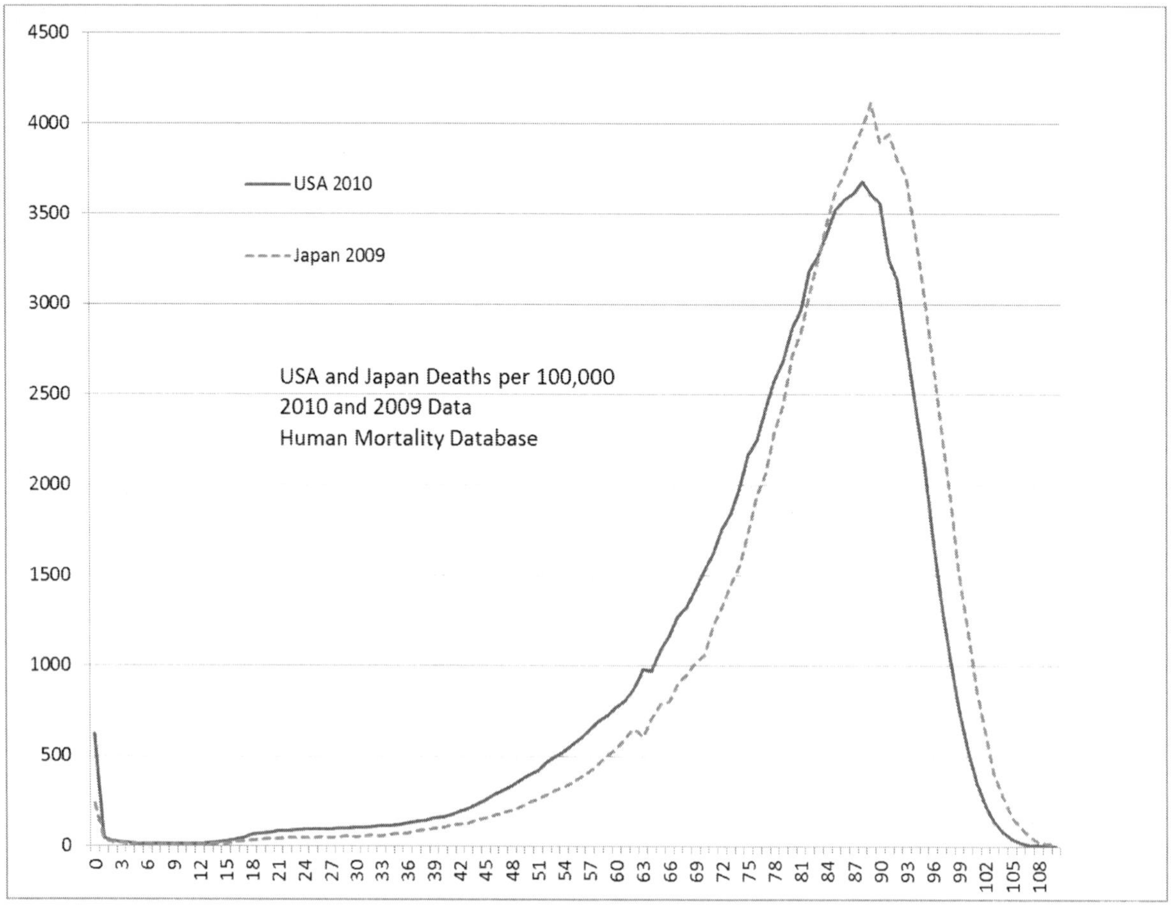

Figure 5 USA and Japan Deaths vs. Age for 2010 and 2009

For ages between 0 to 1 and 15 to 84, Japan has substantially lower death rates!

Aging and Disease

It is obvious from the foregoing that we have been largely successful in conquering diseases that kill people more or less indiscriminately with regard to age or that strike children or young people preferentially. Except for infant mortality, diseases that kill large numbers of people such as heart disease, cancer, and stroke all have incidences that are very highly related to age. Here are the leading causes of death in the total U.S. population:

Cause of Death (USA 2000 Data)	**Percent**
Heart Disease	29.6
Cancer	23.0
Cerebrovascular Diseases	7.0
Lower Respiratory Disease (asthma, COPD)	5.1
Accident	4.1
Diabetes	2.9
Influenza and Pneumonia	2.7
Alzheimer's Disease	2.1
Other	23.5

Many of these diseases are so massively age-related that they are mainly symptoms of aging. For example, your chance of dying from cancer is about 270 *times* greater as an 80-year-old as it is as a 20-year-old. The difference is even larger for heart disease and stroke. Some conditions such as Alzheimer's disease and cataracts are essentially unheard of in young people. Trying to understand the mechanisms responsible for these diseases without understanding the nature of aging seems like a "fool's errand."

Some age-related diseases involve processes that are plausibly a simple function of the passage of time. For example, some heart disease involves buildup of deposits in blood vessels that would appear to gradually increase as time passed. Cancer is thought to involve multiple sequential mutations that would also be increasingly likely as time passed.

However, animals that have much shorter lifespans also develop heart disease and cancer. How is this explained?

Also, there is evidence that even deposits of material causing heart disease can increase or decline based on other factors such as cholesterol levels.

Finally, a disease (See Werner's Syndrome) causing accelerated aging also causes accelerated incidence of heart disease and at least some cancers.

It is therefore clear that many "age-related" diseases are significantly caused by whatever causes aging. Putting it another way, a "major treatment" for aging could reasonably be expected to delay the average onset age of many or even most age-related diseases.

Although a "cure" for many diseases has eluded us, "major treatments" have dramatically increased lifespans of victims of diseases such as diabetes, cancer, heart disease, and even AIDS. We might think of a major treatment for aging as something that would eventually extend average lifespan by 50 percent or more. As we shall see, such a development is no longer in the realm of science fiction.

Aging Variations in Animals

Lifespans vary enormously from species to species. In the wild, animals are subject to attack by predators and competitors, difficulty in finding and competing for food, and adverse environmental conditions such as harsh winters. Older or otherwise weaker animals often do not survive. Under laboratory or zoo conditions, animals have food provided and are protected from other animals of the same or different species as well as from environmental conditions. Zoo animals receive some medical care and typically live much longer than wild animals. The *average* lifespan of wild animals is therefore shorter than for zoo animals while the *maximum* recorded lifespans while usually higher for zoo animals are sometimes higher for wild animals.

Some fish and reptiles have extremely long maximum lifespans and will be further discussed in Chapter 6.

Humans have the greatest authenticated maximum lifespan (122 years) of any mammal. However, some researchers think that a bowhead whale killed in Alaska by Eskimos was 211 years old.

The following table lists *maximum recorded* lifespans (in years) for some animals.

Species	Age
Laxmann's shrew Sorex caecutiens	2
Human Homo sapiens	122
Highland desert mouse Eligmodontia typus	0.8
Marsupial mouse Antechinus (various)	0.9
Asian elephant Elephas maximus	80

Goldsmith

Little Brown Bat Myotis lucifugus	30
Eastern gray squirrel Sciurus carolinensis	23.5
House canary Serinus canarius	22
American robin Turdus migratorius	12.8
American Crow Corvus brachyrhynchos	14.6
African gray parrot Psittacus erithacus	73
Red-breasted parrot Poicephalus rufiventris	33.4
White-winged crossbill Loxia leucoptera	4
American white pelican	54
Brown pelican Pelecanus occidentalis	31
Beluga sturgeon Huso huso	118
Lake sturgeon Acipenser fulvescens	152
Rockfish Sebastes aleutianus	140
Pygmy Gobi Eviota sigillata	0.2
Pacific ocean perch Sebastes alutus	26
Pink salmon Oncorhynchus gorbuscha	3
Sockeye salmon Oncorhynchus nerka	8
Halibut Hippoglossus vulgaris	90
Aldabra tortoise Geochelone gigantea	152
Wood turtle Clemmys insculpta	60
Eastern box turtle	75
Coahuilan box turtle Terrapene coahuila	9.4

This table seriously underestimates the maximum age potential for longer-lived animals relative to humans. This is because humans have greater medical support and because lifespans of humans are much better monitored. We have measured the lifespans of at least several hundred million people. If we measured several hundred million rockfish what would be their maximum age? Most of the people lived under conditions where they could survive even though seriously weakened by age. What would be the maximum age of rockfish if similarly protected and medically supported?

Evolution Fact and Theory

Evolution theory is very central to scientific theories of aging. Understanding the history and current status of evolution theory is therefore essential in order to understand the history and current status of biological aging theory. In this connection it is very important to recognize that "evolution theory" has two parts that possess different degrees of scientific certainty.

First, we have what could be called the theory of species origin. Did evolution of life on Earth occur? Are humans and other current species descended from earlier, different, and generally simpler species? There is overwhelming scientific evidence supporting this aspect of evolution. Darwin presented much of this evidence in 1859. Since Darwin, a huge mass of additional evidence has accumulated further confirming that evolution of life on Earth has occurred. As a result, there is essentially no *scientific* disagreement with the species origin concept. Evolution is one of the major foundations of biology. We can say that *convergence* has applied to species origin in that our certainty has more or less steadily increased with time.

The second part of the concept put forward by Darwin concerns the *mechanics* of evolution. How does evolution work? What factors are important to the process of evolution? Why have organisms evolved in a certain way? Darwin[2] presented a mechanics concept that included "natural selection" and "natural variation." However, even in Darwin's time, there were significant discrepancies and issues concerning evolutionary mechanics. Since Darwin, scientific discoveries have confirmed that Darwin's mechanics are generally valid but have also (relative recently) disclosed additional issues and questions. As a result, there exists significant scientific disagreement regarding the details of evolutionary mechanics. A number of evolution mechanics concepts have been developed that are essentially additions to or modifications of Darwin's mechanics. In this book, "Darwinian evolution", "traditional mechanics theory", or "orthodox evolutionary mechanics theory" means Darwin's mechanics (along with any obvious inferences). We can also define "orthodox theory" as meaning a belief that natural selection according to the inheritable design characteristics possessed by individual organisms or "survival of the fittest" is the *only* factor that determines the evolved design of an organism. Convergence regarding mechanics has not occurred. Certainty regarding mechanics has decreased for reasons to be described.

The "theoretical" part of evolution has to do with mechanics. The disagreements have a relatively minor effect on general biology but are very significant in connection with aging theories. Aging was one of the original discrepancies.

Finally, we have "derivative work", theories that are based on a particular evolutionary mechanics concept. We will be discussing several theories of aging that are based on Darwinian mechanics as well as several that are based on post-Darwin "alternative" mechanics concepts.

Biological Design

We can consider *design* as referring to all of the aspects of an organism that contribute to the performance of useful functions. A design consists of parts or *elements* that work together to accomplish *functions*. Living things have some unique capabilities. They can reproduce and they can pass information regarding their designs to their descendants. Their descendants can then construct themselves in such a way that they have the same or similar design as their parents. Therefore, duck eggs always produce ducks and goose eggs always produce geese. We can speak of *phenotypic design* as referring to the physical and functional design of an organism, and genetic or *genotypic design* as referring to the underlying inheritable data that defines organism design. Any inheritable phenotypic design difference has a corresponding genetic difference however the reverse is not necessarily true. Some genetic data has no phenotypic effect.

Evolution theory says that the designs of organisms have been accumulatively progressing with the passage of time such that there currently exist organisms having much more complex designs than existed, say, one billion years ago. Design features (evolved characteristics) of organisms can involve incredible *complexity* (more complexity than anything else known to science). Design features of organisms involve *coordination* and *cooperation* between many different tissues and supporting functions. For example, vision in an eagle involves muscles, sensory cells, nerves, brain functions, and inherited behaviors, all of which must be supported by other systems and tissues. The physical or anatomical design of living organisms is easily observed. The chemical elements and their interrelationships are less obvious but probably generally more important.

Goldsmith

The most important question to be discussed in this book is: Are living organisms *designed* to age and therefore genetically *programmed to age*? If so there are consequences. If aging is a function like vision we can expect it to have characteristics common to other evolved functions including complexity and coordination. As this book will show, modern evolutionary mechanics theories and recent discoveries overwhelmingly support programmed aging.

2. Evolutionary Mechanics Theory

Figure 6 Charles Darwin

Scientific theories regarding the cause of aging are intimately bonded to theories of evolution, particularly those of Charles Darwin. While there is no current scientific disagreement with the idea that evolution of life on Earth has occurred there is still significant scientific disagreement regarding the *mechanics* of the evolution process. How does evolution work? What aspects of living organisms are important to the evolution process? These disagreements are crucial to modern aging theories.

Charles Darwin (1809 – 1882) was born in Shrewsbury, England to a family with long experience in medicine, and studied medicine at Edinburgh, Scotland. Darwin found medicine boring, and disliked the sight of blood, and so eventually became interested in the study of plants and animals. At one time, he spent several years studying barnacles. Eventually he sailed as naturalist on the HMS *Beagle* during a survey mission to South America from 1831 to 1836.

Darwin's book, initially published in 1859, *The Origin of Species by Means of Natural Selection or, The Preservation of Favoured Races in the Struggle for Life* was a best seller and immediately controversial, both because of scientific disagreement and religious issues. Darwin's primary conclusion, that species are descended from other species, was a lightning rod for controversy. Another English naturalist and contemporary of Darwin named Alfred Wallace (1823 – 1913) had a similar theory and is considered by some to be a co-discoverer of the evolution theory, though much less well known.

It is difficult to overstate the impact Darwin had on biology. In the 150 years since, there is no other theory in the field of evolution that has anywhere near the scientific acceptance of Darwin's theory. Popularly, most adult Americans easily associate "Darwin" with "evolution" and vice versa. Darwin's books are still carried by most bookstores.

Although most people think of "Darwin's *Theory* of Evolution", it is important to note that Darwin actually proposed *two* theories: The first, the idea that species were descended from other species, we could refer to as "Darwin's Theory of Species Origin." The second, in which he proposes how evolution works, and what factors are important to or influence the evolution process, we could refer to as "Darwin's Theory of Evolutionary Mechanics." It is this second theory that is critical to various theories of aging.

Prior to the theory of natural selection, there was little or no reason for naturalists to distinguish between aging and any other species-specific characteristic of an animal. Whatever caused a rat to have beady eyes, curved teeth, and a long tail, presumably also caused it to have a certain lifespan.

If we compare aging with other animal characteristics such as the type and size of teeth we find that teeth vary somewhat between individuals of a given species. Aging also varies somewhat between individuals.

Goldsmith

Teeth characteristics vary to a greater degree between species. Aging varies greatly between species.

Teeth characteristics are inheritable. Aging is also somewhat inheritable. If your ancestors lived long lives, you are also likely to live a long life.

Teeth characteristics seem to fit reasonably well with the other characteristics of an animal. Small animals have smaller teeth. Carnivores and herbivores have teeth suited to their respective diets. Similarly, longevity seemed to fit the development and reproductive life cycle of an animal. Animals that developed to maturity rapidly and could breed rapidly tended to have shorter lifespans. Animals that needed a longer time to develop and become sexually mature tended to have longer lifespans.

Some species had bizarre examples of teeth. For example, boars have specialized teeth that extend outside the mouth in the form of tusks. Some deep-sea fish seem to consist mainly of teeth! Similarly, there are bizarre examples of aging such as the salmon, which lives for several years in the sea, swims up a stream to spawn, and then dies almost immediately from a tremendously accelerated aging process.

Darwin and the theory of natural selection resulted in lifespan being considered a substantially different type of biological property from teeth and most other animal design characteristics.

The Theory of Natural Selection

Darwin proposed that random changes occurred in the inheritable designs of organisms. Through the process of natural selection, the changes that produced a beneficial result were retained in subsequent generations of the organism. Changes that were adverse were rejected.

Many of Darwin's conclusions reached in *Origin* were based on comparison of the generic characteristics of wild plant and animal species with those of domesticated plants and animals. Darwin was especially interested in the *differences* or *variations* between individual members of species and between different species. The wild species were the result of *natural selection* or what was later called "survival of the fittest" while the domestic species were partly the result of selective breeding by humans. Wild species were subject to predator attack, intra-species warfare, starvation, and exposure to adverse weather. In contrast, domesticated species are protected and provided with food. Wild species survived based on survival traits. Domesticated species were bred for specific qualities desired by human breeders. Most domesticated species would not survive long in the wild. Domesticated species were frequently bred for externally obvious characteristics where natural selection would affect any characteristic, internal or external, if it affected survival. Darwin noted that the differences between members of domestic species such as dogs were grossly larger than the differences between members of wild species. Domestic species represented a proof that "selection", that is selecting successive generations for the same property, could eventually result in very large changes to organisms.

Darwin determined that most characteristics or *traits* of organisms were *evolved adaptations* that aided the organism in surviving. That is, if sharper teeth resulted in animals of a given species surviving longer and therefore having more descendants than other animals with dull teeth, eventually, most of the animals in the population would have sharper teeth because the genetic design for sharper teeth would be passed to the larger number of descendants. Sharper teeth resulted from the process of evolution, the process of *adapting* to conditions in the animal's external world.

The *mechanics* or how-it-works aspect of Darwin's theory of natural selection is very simple and readily understood. There is no need for advanced mathematics or other major complexity. High school students can readily understand survival of the fittest.

Even the slowest reproducing species, would, if allowed to breed in an uncontrolled manner, occupy the entire planet in a relatively short period of time (an idea earlier put forth by Thomas Malthus (1766 – 1834)). This is the idea of geometric progression where one rabbit has two progeny, and they each have two, and they each have two, etc.

The growth of the populations of all species is therefore *checked* by a variety of external factors such as predators, diseases, food supply, and environmental conditions so that in a stable population each individual has an average of only one descendent which survives to produce one descendent and so on. All wild organisms are in competition for survival with other species and even more acutely with members of their own species.

A major clue as to the nature of evolution was found in the *geographical distribution* of various plants and animals. Darwin's expeditions provided very extensive data in this area and *Origin* has two chapters on geographical distribution.

If an inheritable beneficial change occurred in an organism, we would expect its descendants possessing that change to propagate radially (or *radiate*) from the point of origin. Natural barriers such as oceans would tend to prevent the free migration of land animals and many plants. If species were descended from other species, we would expect substantial differences between such descendent species that were separated by such a barrier. "Families" of species that lived in one area would tend to resemble each other more than they resembled species in another area just as individuals in a geographic area resemble each other more than they resemble individuals that lived in a distant place. Sure enough, Darwin found flora and fauna to be "utterly dissimilar" between barrier-isolated geographic areas that were otherwise similar such as Australia, Western South America, and South Africa. Areas that had greater difference in their climates or otherwise represented different habitats but were not separated by migration barriers tended to have more similar organisms.

Darwin recognized that changes to inheritable characteristics that had no particular survival advantage or disadvantage, (survival neutral characteristics), could also geographically propagate and eventually become widespread.

Variation and Incremental Steps

Darwin proposed that small random changes to the inheritable characteristics of organisms (*mutations*) were the driving force behind evolution.

The probability that an animal will live longer and have more descendants is determined by the *combination* of all the characteristics possessed by that animal. For example, suppose some herbivore species in Africa becomes heavily predated by lions. We can suppose that "faster" might be a desirable characteristic for these animals. However, "faster" is actually the result of many characteristics. If we were redesigning this animal for more speed we might give it longer leg bones and reduce the size of other bones and systems to save weight. We might want to reduce heavy defensive characteristics such as horns. Internally, we would need to increase the size of some muscles and joints. Nerves, brain, and inherited behavior patterns would all need modifications.

All these changes are individually adverse and only have full survival benefit when applied in a specific combination. Larger leg bones without larger muscles would be a disadvantage. Larger muscles without stronger ligaments, tendons, and joints would be a disadvantage. A mutation that caused a large change in leg bone size would be adverse because it would not be accompanied by the other necessary complementary changes that are required in order to result in a beneficial effect.

Darwin was of course aware that occasional mutations caused large changes in animals. Darwin referred to these as "monstrosities" because such mutations (such as two-headed animals) were adverse and ugly.

Therefore, *beneficial mutations* in more complex organisms would be confined to those that caused only minor changes. A small increase in leg bone size might be beneficial. Subsequent mutations that resulted in small changes in associated muscles, tendons, or nerves might also be beneficial. Yet subsequent small additional increases in bone size might now be beneficial because the tendons and muscles had developed to match the original change in bone size. Eventually, in tiny increments, an animal with dramatically larger legs might evolve. If we examine the differences that currently exist between normal members of a wild species, they are indeed relatively minor in terms of their survival effect. We can use the term *propagatable mutation* to describe a mutation that had a sufficiently small effect that it could plausibly spread in the population. Such a mutation might have no physiological effect or might even have a very small negative effect.

This example also illustrates that design characteristics of organisms involve *tradeoffs*. In this case, we are giving up heavy defensive features such as horns, long sharp claws, and strong jaws in favor of saving weight in order to be faster.

Variation in inheritable characteristics between individuals in a population is thus a required property of life to support Darwin's evolution theory. Some contemporaries thought new species were created instantaneously by massive mutation in a single individual. Natural selection operates upon this variation. Our increasing knowledge of the details of inheritance (see Genetics) provides increased support regarding the essential evolutionary role of variation.

In connection with variation, Darwin recognized that *genetic diversity* was a benefit to survival. Highly inbred domestic animals and plants were generally weaker, less hardy, and more susceptible to disease than *crosses* between more diverse specimens.

Darwin concluded for reasons described above that evolution took place by means of tiny increments. Each generation was only minutely different from its parents. Webster's defines "evolution" as "...gradual development."

These two features, evolution by means of natural variation and evolution in tiny incremental steps were the center of Darwin's mechanics theory. When Darwin referred to "my theory", he was referring most specifically to these features in addition to the idea that species evolved from other species in the same manner.

In addition to physical characteristics, Darwin included instincts and inherited behavior patterns in traits that evolved through natural selection. Behaviors would need to evolve in parallel with physical characteristics. A wing has no survival value unless used to fly. An eye is useless unless used to see. Flying and seeing would need to be supported by the appropriate brain and nervous system characteristics including inherited behaviors that lead to learning to fly and learning to see. Even a light sensitive spot on a worm would have no added survival value unless it somehow altered the behavior of the worm.

In connection with evolution we need to discuss some aspects of probability theory because whether an individual organism lives longer or breeds more is a matter of chance. Imagine that we had a quantity of coins that mixed those of two different types. When "flipped" one type has a 50 percent probability of landing "heads" (and 50 percent probability of landing "tails." The other type has a 60 percent probability of heads (40 percent tails). Suppose the coins are otherwise indistinguishable and we desired to separate the two types. We could run a number of flipping *trials* on each coin and note the results (heads or tails). The results of one trial are not significant (it could go either way). However by making a number of trials on each coin we could definitively distinguish between the types. Now suppose that the second type of coin had a 50.25 percent chance of heads and 49.75 percent chance of tails. We could still definitively distinguish between the two types but it would require a much larger number of trials on each coin. The number of trials required is inversely proportional to the size of the probability difference. In the evolution process, the life of an organism is essentially a trial of the particular combination of design properties possessed by that organism. Darwin's "tiny steps" concept means that a large number of trials (organism lives) would be necessary to distinguish between organisms possessing relatively small differences in their survival probabilities in the natural selection process.

Another observation is that in complex organisms such as mammals the inheritance system is generally capable of independently specifying and evolving the characteristics of a very large number of design elements. If an animal needed longer toes on the same size foot, evolution could do that. If an animal needed a bigger foot but smaller toes, nature could do that too. Even if we consider only mammals, there are examples of essentially any possible combination of foot parameters. This observation is important in evaluating the plausibility of some theories of aging (to be described) that hold that aging is an unavoidable side effect of a beneficial property.

Some important implicit requirements of the theory of natural selection should be mentioned.

First, the natural variations in characteristics in a population of animals must be genetically programmed and thereby inheritable. Variations that were not genetically recorded could not participate in evolution because they would not affect the genetic content and therefore the design of subsequent generations. Note also that natural selection requires *local* variation. It is irrelevant if the squirrels on the other side of the planet are faster or smarter than those on this side. The variation must exist between individuals that could plausibly be in competition.

Second, evolution requires a population. Evolution results from the *difference* in statistical lifespans between animals that have a beneficial trait and those that do not.

Third, in order to evolve, a trait must be *expressed* or displayed by the organism in such a way that it affects the differential in lifespan or reproduction. The trait must cause a functional *performance* difference. A *latent characteristic* which was present but not expressed at the time of an animal's death could not have affected whether it lived or died. The death of that animal therefore could not have contributed to evolution of that characteristic.

Finally, and very important, the *probability* that an individual animal would live longer and/or breed more is determined by the *combined effect* of all its characteristics.

As you will see later in this book, these requirements are central to discriminating between various theories of aging.

Few scientists of the time would have argued against the idea that natural selection could cause a species to evolve. After all, humans had for thousands of years been causing domesticated species to change by selective breeding. If small dogs were bred with small dogs for a long time, men could produce a Chihuahua. If fast dogs were bred with fast dogs for long enough a Greyhound would result. If you go back far enough, Chihuahua and Greyhound are descended from the same dog. Selective breeding could produce an organism design that did not previously exist. "Natural" selection could obviously cause the same sorts of changes in the design of a species.

The argument was whether all the *species* that now exist could have evolved from a single original primordial species (probably a single-cell organism on the order of pond scum) simply by the effects of natural selection acting in slow small increments on individual variation caused by random mutation. It was a much greater leap to believe that humans evolved from pond scum than to believe that fast dogs developed from slow dogs. It appeared to be clear that no amount of selective breeding of a dog could ever result in producing a cat. Selective breeding (or natural selection) could only enhance or reduce characteristics that *varied* between individual members of the species. Properties of a species that varied between species but did *not* vary between individual members could not be affected by selective breeding or natural selection. Darwin's central idea was that a mutation would occasionally alter these species-invariant properties in a single individual and then propagate to its descendants, essentially adding to the properties that varied between individuals. This new property could then be affected by natural selection (or selective breeding).

Development of different varieties was driven by geographic separation and differences in conditions. If a mammal lived in an area that contained both mountains and lowlands, it could not be optimized for both areas. The animals in the mountains might tend to develop characteristics that would favor survival in the colder, higher areas such as increased fur. The lowlands animals might develop characteristics favoring their conditions. Since they were somewhat geographically isolated the two groups would tend not to interbreed (genetic isolation). The two new varieties are both more effective at surviving in their particular habitats than the original variety and so would probably replace the original variety. After a long time the two varieties could evolve to be so different, they would become separate species. Eventually, the number of new species and variations being produced would be more or less matched by old species becoming extinct.

Besides natural selection, which is based on survival, Darwin recognized that *sexual selection* also played a role in evolution. Sexual selection would involve advantages that an individual might have that did not affect its survival but did represent an increase in its probability of breeding such as ability to attract the opposite sex. Darwin considered that sexual selection was weaker than natural selection.

Humans would be classified in Darwinian terms as "domesticated" as opposed to "wild" animals because humans have probably not existed under wild conditions for thousands of years. For example, one would expect the incidence of genetic diseases among humans to be higher than in a wild species because the effects of civilization and medical intervention allow individuals with adverse inherited conditions to survive and propagate in a way not possible in a wild species. Each major medical advance that allows people to survive and reproduce where formerly they would have been more likely to die without reproducing contributes to weakening the human genome. We are increasingly dependent on medicine, other technology, and societal advances to compensate for a deteriorating genome. A case can be made that consequently humans are no longer genetically evolving but rather genetically devolving while simultaneously "evolving" in the development of compensating knowledge that is non-genetically transferred between generations. Darwin speaks to such aspects of evolution of humans in his later book *Descent of Man* (1871). Use of human data (such as actuarial data) in an attempt to prove or disprove theories based on natural selection (such as aging theories) is therefore highly suspect although commonly done.

In connection with "survival of the fittest", the term *fitness* came to be associated with the ability of an organism to survive *and* breed. It was apparent that there could be compromises or tradeoffs between various traits that improved survival. Speed in an animal might be a compromise with strength. A faster but weaker animal might be able to survive better than a slower stronger animal. It was also apparent that a tradeoff could exist between survival and reproduction. An animal that was a prolific breeder might be as fit as an animal with better survival capabilities but reduced breeding capabilities. This particular tradeoff is important to some aging theories. The fitness concept does *not* incorporate any aspect leading to increasing the *quality* of progeny relative to their parents. The quality of progeny is assumed to come from the effects of natural selection on the parents. Most biologists define fitness along the lines of "the ability to produce adult progeny." Note that survival, per se, does not convey an evolutionary advantage. The evolutionary benefit of survival is in increasing opportunity to reproduce.

Darwin's mechanics theory has some other important properties. Evolution does not take place *during* an organism's life as suggested by some earlier theorists. The inheritable beneficial characteristics of an organism are *fixed* during its life.

Evolution is a very slow process. Although mutations might occur occasionally, *beneficial* mutations might be very rare because most mutations could be expected to be adverse. However, the *beneficial effect* caused by a beneficial mutation, once it finally happened, would be immediate. The immediate descendants of the mutated organism would live longer and breed more. The immediate "short-term" effect of beneficial mutations in Darwin's concept is important for theory considerations to be discussed.

Darwinian Evolution Requires Individual Benefit

Many people understand that Darwin proposed that organisms evolve design characteristics that benefit the organism but have a somewhat philosophical view of what constitutes "benefit." Perhaps "benefit" includes benefits to the species, population, herd, "greater good", etc. It is therefore important to note that because of the mechanics of Darwinian evolution (survival of the fittest and the short-term nature of Darwinian mechanics), in orthodox mechanics theory, beneficial mutations are associated with *individual* organisms. Beneficial genetic mutations propagate because the *individual organisms possessing those mutations* can live longer and breed more and therefore better propagate their personal designs in the population. We can speak of "individual" fitness and "individual" benefit to emphasize the importance of the individual in orthodox Darwinian evolutionary mechanics. Specifically, traditional evolution theory does not allow for the evolution of traits that are beneficial to "the herd", or "the group", or "the species" if they are adverse to individuals. This ***individual benefit requirement*** will become very important in discussions to follow.

After original publication of *Origin*, there was some popular confusion about the term "natural selection." Some people were interpreting "selection" as meaning a function of God. Subsequently, the expression "survival of the fittest" was used by Darwin and others in an effort to clarify that organisms themselves were actually performing the "selection."

Darwin's main conclusion, that species are descended from other species, is supported by overwhelming and growing scientific evidence. As will be described, there remain scientific disagreements regarding Darwin's mechanics theory.

It should be noted that until the nineteenth century, there was little scientific evidence that conflicted greatly with the biblical notion of creation, that is, that the Earth and all the species on it had been created more or less simultaneously in the relatively recent past. It was widely thought that the Earth was not very old, possibly as little as 25,000 years old.

For example, mountains could not be extremely old. Every wind that blows and every rain that falls removes material from a mountain and deposits it in a valley. If the Earth were extremely old, would it not be essentially flat?

The discovery of the occasional odd bone was often attributed to existing but undiscovered species or recently deceased species. There was little evidence that existing species had changed much or new species appeared during the time that people had been making observations, which was a significant fraction of the putative age of the Earth.

However, eventually it became clear that the Earth was *very* old, on the order of 4.5 *billion* years old. Mountains and other geological features were being replenished by geological processes such as plate tectonics that, even now, seem fantastic. Methods were developed (eventually including radioisotope dating) for estimating ages of rocks and fossils, which disclosed extreme ages for some fossils and allowed the determination of a time continuum for the appearance and disappearance of various life forms. All of these developments contributed support to the idea that evolution of life on Earth had in fact occurred.

There was great and immediate objection to Darwin's theories on religious grounds. This controversy greatly contributed to the popularity of and very wide distribution of Darwin's books. The major objection was of course Darwin's idea that species were descended from other species as opposed to being individually created by God. The idea that the human species was descended from "monkeys" as opposed to being individually created by God was particularly unattractive.

In 1925 Tennessee passed a law "...prohibiting the teaching of the Evolution Theory in all the Universities, Normals and all other public schools of Tennessee, which are supported in whole or in part by the public school funds of the State, and to provide penalties for the violations thereof."

Later in 1925, a high-school biology teacher, John Scopes, was charged with illegally teaching the theory of evolution. The subsequent trial, State v. John Scopes, known popularly as "the monkey trial", pitted three-time presidential candidate William Jennings Bryan as fundamentalist council for the prosecution against famed agnostic lawyer Clarence Darrow for the defense and created an international media circus in the little (pop. 1,800) town of Dayton Tennessee. The trial was eventually transferred into a tent so that thousands of spectators could be accommodated. Bryan also acted as a prosecution witness and was cross-examined by Darrow in a famous exchange, which, by all accounts, Darrow won handily.

Scopes was obviously guilty. The trial was about the constitutionality and reasonableness of the law. Darrow actually requested the jury to find Scopes guilty so the case could be appealed to the Tennessee Supreme Court, which eventually dismissed the case. The international derision resulting from this case inhibited many other states that had been considering anti-evolution laws. Eventually, in 1968, the U.S. Supreme Court found laws prohibiting teaching of evolution in public schools unconstitutional by virtue of the First Amendment (separation of church and state).

Fundamentalist anti-evolution efforts continue today. Creationists favor teaching a biblical version of the creation in public school science classes as an alternative to evolution theory. Creationist texts abound with scientific-sounding arguments, footnotes, and references. However, unless a dramatic rightward shift in the U.S. Supreme Court occurs, teaching the Bible in U.S. public school science classes will remain unconstitutional. Creationists attack the species descendancy part of Darwin's idea and are therefore in the position of opposing the overwhelming scientific evidence supporting species descendancy.

Intelligent Design (ID) holds that individual species of living organisms, because of their complexity and for various other reasons, cannot have arisen from random chance and natural selection, and therefore must be the result of the operation of some intelligence. ID proponents avoid mention of God or the Bible. Readers are free to ascribe the intelligence driving the development of different species to little green men, Klingons, or whatever their favorite source of supernatural intelligence might be. ID proponents tend to emphasize "random" mutations and largely ignore the incremental and accumulative effect of natural selection in their arguments that complex life forms could not have been produced by natural forces. Credentialed scientists approach rural local public school boards in religiously conservative states with arguments to the effect that ID represents a legitimate scientific disagreement with traditional evolutionary mechanics theory and should therefore be taught as an alternative.

ID has an advantage over creationism in that ID proponents can choose to believe that the historical record matches that proposed by evolutionists, that various species appeared on the same schedule, and that individual species are derived from prior species. The only difference is now that God or other source of supernatural intelligence and power directed and implemented the design of each individual species. The fossils and other evidence therefore do not conflict with ID as they do with creationism.

However, ID and creationism are actually fundamentally incompatible with science. The development of a "scientific" theory becomes trivial if the theorist is free to invoke God or other source of supernatural direction anytime he is having difficulty making his theory agree with observed facts. Once it was determined that God was responsible for a certain function or process, further inquiry would be inhibited or could even be prohibited by clerics. If, several hundred years ago it had been determined that God was responsible for lightning, would we ever have discovered and harnessed electricity?

Many or even most scientists have at least some belief in a deity. Many have some belief in their personal supernatural ability to affect the laws of probability (useful for gambling). Others observe various superstitions. However it is understood that the study of *natural* science (like biology) specifically excludes all *supernatural* factors. Teaching religious beliefs about evolution is appropriate in comparative religion class or sociology class but not in biology class.

Many people are not as opposed to the *fact* of evolution as to the *teaching* of evolution. They see teaching evolution, especially in lower grades, as anti-religion, essentially teaching atheism. While the "humans are descended from lesser species" aspect of Darwinism is obviously objectionable to fundamentalists, another aspect, the individual benefit requirement, which leads to a "dog-eat-dog" view of the natural world, is more generally objectionable. Most religions, societies, and civilization generally, are built on the concept of individual sacrifice for the greater good. Horrible acts including pogroms and "ethnic cleansing" have been "justified" based on evolution theory.

Anti-evolution efforts in the United States are having a significant effect. A Harris poll in June 2005 found that 54 percent of Americans do not believe that humans developed from earlier species (up from 46 percent in March 1994).

The existence of creationism and ID contribute to a sort of scientific backlash, a "siege mentality", an atmosphere of "us versus them" in the scientific community. Legitimate scientists feel comfortable in taking positions that attribute more certainty, scope, and comprehensiveness to traditional Darwinian mechanics theory than is actually scientifically justified. See example in chapter 7. This is an important *non-science factor* that influences thinking about evolutionary mechanics and therefore influences aging theory.

Miscellaneous Objections

In later editions of *Origin*, Darwin provided a chapter, *Miscellaneous Objections to the Theory of Natural Selection*, in which he responds to objections raised by contemporary scientists.

If Darwin's theories were correct, there would have existed sometime in the past individuals possessing all of those little incremental variations extending from that original organism to each current species. One objection was that the fossil record did not seem to support this, as there were times when new species seemed to suddenly appear and there were many "missing links." Darwin presented extensive geological arguments showing that the geological fossil record itself had gaps that would explain the non-discovery of some intermediate forms. For example, a geological event that caused a coastal area to submerge might cause loss of some part of the fossil record.

Another major objection had to do with the intermediate survival value of various organs and structures. Contemporaries argued that "half a wing" would not be of any value and so a full wing could not have evolved by means of tiny steps. Darwin agreed that his theory would be defeated if a single case could be found where an organ or structure did not have increased survival value in all of the incremental intermediate forms needed to evolve that organ or structure.

For example, the eye is a complex structure that might not appear to have value without all its complex parts including retina, lens, iris, etc. If this were true, there would be no incremental evolutionary path from no eye to complete eye. Darwin was able to show that even in existing living species, there are examples of a continuum of minor variations of optical organs all the way from a light sensitive spot on a worm to an eagle's eye. He was also able to show that eyes in current animals evolved down at least two different evolutionary paths.

However, although the theory of natural selection explained a great many things, and Darwin was able to successfully respond to many objections (such as those above) from contemporary critics, there remained some areas where Darwin was unable to explain observed discrepancies. For example, colony insects such as bees and ants have types of individuals such as workers or warriors that are sterile. Since they are sterile, their survival or non-survival could not determine whether beneficial traits were propagated and natural selection, rigidly interpreted, did not apply. The warriors and workers had zero fitness because they could not breed. Darwin's response in this and some similar cases was that beneficial characteristics of the warriors and workers benefited their *species* although the exact mechanism whereby evolution actually occurred was unclear. Some animal behaviors, (to be discussed later), also did not seem to fit Darwin's theory.

Darwin's Dilemma

Another major contemporary objection, central to the subject of this book, was the relative absence of longevity. It was immediately apparent that observed animal characteristics regarding aging and longevity did not fit the rules set forth by Darwin for natural selection. Contemporary scientists cited aging and longevity as a demonstration that Darwin's theory of natural selection was incorrect using the following logic: Since longevity was of value in increasing the survival time and breeding opportunity of any organism, would not natural selection (if true) result in ever-increasing longevity? Wouldn't aging, since it was obviously adverse to fitness be "selected out" by the process of natural selection? In other words, Darwin's theory predicts that animals and humans should not age.

Darwin's answer was as follows:

> "A critic has lately insisted, with some parade of mathematical accuracy, that longevity is a great advantage to all species, so that he who believes in natural selection

> "must arrange his genealogical tree" in such a manner that all the descendants have longer lives than their progenitors! Cannot our critics conceive that a biennial plant or one of the lower animals might range into a cold climate and perish there every winter; and yet, owing to advantages gained through natural selection, survive from year to year by means of its seeds or ova? Mr. E. Ray Lankester has recently discussed this subject, and he concludes, as far as its extreme complexity allows him to form a judgment, that longevity is generally related to the standard of each species in the scale of organisation, as well as to the amount of expenditure in reproduction and in general activity. And these conditions have, it is probable, been largely determined through natural selection."

It is clear from Darwin's response that he believed that lifespan was a characteristic determined by natural selection, that is, an *evolved characteristic* or *adaptation*. The observed limits on lifespan were the result of some internal (design) factor as opposed to being externally imposed because similar species were observed to have different lifespans. Darwin also believed that at least in the case of species with a shorter lifespan that a limited lifespan somehow provided compensating benefit.

However, Darwin could not explain *how* a limited lifespan benefited members of any particular species or *why* that benefit outweighed the disadvantage of a shorter lifespan. His argument is therefore circular: My theory says there must be some hidden compensating (individual, theory conforming) benefit so therefore there must be one. The theory is being used to predict the observation as opposed to observations resulting in a theory. Note that this same "explanation" could be used to "explain" *any* instance of an apparently individually adverse organism design characteristic.

At the time this was a reasonable position. There were perhaps thousands of lifespan observations that appeared to conflict with the individual benefit requirement. At the same time there were millions of observations of plant and animal design characteristics that clearly conformed to Darwin's ideas in that they obviously aided the organism in surviving or reproducing. Darwin had a reasonable expectation that eventually compensating individual benefits would be discovered for the small minority of apparently non-conforming observations.

You will see variations of this circular argument made by other theorists discussed later in this book.

Of course, what was a reasonable position in 1859 is not so reasonable today. By now, certainly we should have been able to definitively identify the individually beneficial compensating effect. And yet, in many cases, only Darwin's hope for eventual discovery remains. In addition, other apparent conflicts with the individual benefit requirement have surfaced.

"Darwin's dilemma" has been a major constraint to every subsequent attempt to devise a scientific explanation for aging for nearly 150 years. Scientists have been forced to choose between believing that aging was an adaptation, a deliberate part of organism design, despite orthodox mechanics theory, and believing that it was not an adaptation despite the massive and growing observational evidence that it was.

Reproductive capacity as a function of calendar age has issues very similar to those of aging. An evolved design characteristic that purposely caused a decline in reproductive capacity with calendar age would appear to be obviously individually adverse and therefore incompatible with traditional theory. Fundamental limitations that result in such a decline have issues similar to aging regarding the wide range of observed differences between species.

Summary of Darwin's Theories

-Main Conclusion: Species are descended from other, earlier species.

-Random mutations that cause beneficial inheritable changes are incorporated into an organism's design by means of natural selection. Adverse changes are rejected.

-Living organisms can adapt to changes in their environments by means of natural selection.

-Species and individual organisms are in competition for available resources.

-Evolved characteristics (adaptations) of organisms are determined *entirely* by natural selection (survival of the fittest). Evolved characteristics exist because they beneficially affected an *individual* organism's ability to survive or breed.

-Evolution is a slow, incremental, process.

-Genetically transmittable natural variation between individual members of a species is essential to the evolutionary process.

-The theory is generally applicable to all living organisms.

Unresolved Discrepancies with Darwin's Theory

One of the troublesome aspects of Darwin's theory was its emphasis on *individual* survival and fitness, which leads to a brutal, "red of tooth and claw" view of the biological world. Human society and civilization are built on *cooperation*. Although some forms of cooperation are individually beneficial, societal rules, the Ten Commandments, criminal and civil law, and so forth, are largely based on *restricting individual behavior* in favor of the group.

Of course, someone might say that humans are fundamentally different from animals. This view is internally inconsistent with Darwinian evolution theory in that a main point of Darwin's theory is that humans are not so different from other animals and that human design resulted from the same processes that determined the designs of other organisms.

In addition to the colony insects that exhibit grossly "societal" behavior at the expense of individuals, there are many observations of animals that seem to disclose societal, individually adverse behavior.

This conflict between observed behaviors or other characteristics that benefit the group at the expense of individuals and Darwin's theory is a recurring issue to be discussed at length.

Since Darwin, we have vastly increased our understanding of the mechanics of inheritance. This information, essentially the whole science of genetics, also discloses apparent problems and disagreements with details of Darwin's mechanics theory.

Here is a summary of apparent unresolved discrepancies (discussed in this book) between Darwin's mechanics and various observations:

-Gradual aging

-Biological suicide

-Some societal behaviors

-Altruism

-Apparent unnecessary delay in reproductive maturity (especially male) in some species

-Some mating rituals and other aspects of sexual selection

-Sexual reproduction (massively individually adverse)

-Many aspects of genetics

All of these discrepancies conflict with orthodox Darwinism's individual benefit requirement.

Altruism refers to observations of animal behaviors that are not in the best interest of the animal from an orthodox theory "dog-eat-dog" viewpoint. Example: an animal risking its own life protecting unrelated young of its species. According to traditional theory an animal should protect its mate and direct descendants but not unrelated animals with which it is in competition. Although altruism seems to be a relatively minor discrepancy with orthodox theory, for some reason it has historically attracted the most interest in developing alternative evolutionary mechanics concepts.

Sexual reproduction, which evolved from earlier asexual reproduction methods, is massively individually adverse for several reasons: First males are relatively reproductively useless. Imagine some turtle species. The turtles mate and subsequently the female lays eggs and walks away. Neither parent provides any nurturing or protection to their young. The size (developmental maturity at hatching) and number of eggs the female can produce is limited by physical constraints within the female's body. It is obvious that if a variety of these turtles could produce twice as many eggs (as they could if males also produced eggs) that such a change would represent a large increase in individual fitness. Sexual reproduction therefore represents a massive, factor of two, decrease in reproductive fitness relative to asexual reproduction.

Second, sexual reproduction dilutes the ability of an organism to transmit its own personal design. It cannot be sure that its personal design will be reproduced in descendants as it would be if reproducing by asexual reproduction similar to cloning. If the evolutionary goal of all organisms is to propagate their individual designs, sexual reproduction would appear to be a massive step backward.

If an individual happened to be isolated from any members of the opposite sex of its species during its mating season it could not reproduce, a problem that would not exist if it could reproduce asexually. Finding a mate entails risk. The individual might die before finding a mate.

So why don't the turtles reproduce asexually? What individual benefit compensates for the massive individual disadvantage? Alternative evolutionary mechanics theories suggest an answer. See discussion of evolvability theory later in this book for one explanation.

Male puberty age in many animals is delayed relative to the age at which it would be expected to occur from a strictly developmental (growth) standpoint. Puberty in humans resulting from abnormal conditions has been observed at ages as young as four. In mammals or other species in which the male provides a nurturing or protection function, one could suggest that there was individual benefit in delaying puberty such that the male parent would be better able to perform that function. However in animals such as the turtle, delayed male puberty is an individual disadvantage.

Evolutionary Philosophy

Everyone reading this book already has some beliefs regarding the process of biological evolution. These beliefs have likely been acquired and retained during an extended period and may be unlikely to be altered by what you read here. As we will see, beliefs and philosophical approach regarding the evolution process logically and understandably almost entirely determine one's thinking regarding "evolutionary" theories of biological aging.

Which of the following most closely resembles ***your*** beliefs regarding the mechanics of evolution?

1. The evolution process is directed *entirely* by "survival of the fittest" and *individual* benefit. We completely understand the evolution process; there is nothing left to know. It is impossible, 150+ years after Darwin, that there could be other natural factors involved in determining the direction of the evolution process. Any apparent conflicts that may be observed must be the result of experimenter error, fraud, misinterpretation of the data, or must otherwise be explainable in ways that are compatible with the above existing understanding of the evolution process. Efforts should therefore be expended in finding explanations for observations that fit the theory as opposed to developing a new or modified mechanics theory that fits the observations. Conflicting observations should be discounted.
2. Darwinian survival of the fittest is the most important factor directing the evolution process but it is likely that eventually science will discover and generally accept other *natural* factors that influence evolution. The persistence and growth of credibly observed inconsistencies and logical issues demands we consider this. Most things in biology (even science in general) have eventually turned out to be more complicated than they originally appeared. Some or all of the existing apparent conflicts could be explained by the new discoveries and determinations. Observational evidence provided by multiple trusted observers should therefore be accepted on its face value despite conflict with orthodox theory. Efforts should be expended in developing a modified evolutionary mechanics theory that best fits all of the current credible observational evidence.
3. Survival of the fittest is the most important factor influencing the evolution process but other natural factors *do* also influence the process. Specifically *group benefit* or *benefit to the evolution process* can influence evolution in addition to individual benefit. We should look for explanations for observations that include those suggested by these new evolutionary mechanics theories.
4. None of the above. *Super*natural factors are involved in directing the evolution process or evolution of life on Earth never occurred.

If you believe (1), you believe in traditional evolutionary mechanics. If you believe (2) you are an evolutionary mechanics "agnostic." If you believe (3) then you are a believer in an *alternative* evolutionary mechanics theory (to be described in detail). If you believe (4) you are a creationist or Intelligent Design proponent and might be comforted by the fact that the *natural* science community cannot agree on which of the first three philosophies regarding evolution mechanics theory is correct!

It should be obvious that evolutionary mechanics science is relatively "soft" science compared to other areas of biology much less most other areas of science. Many aspects cannot be experimentally tested. Evolutionary mechanics is consequently relatively heavy on "philosophy" and light on "science." As a result, *non-science factors*, to be discussed, are much more important in influencing thinking about evolutionary mechanics than they would be in a "hard" science discipline where repeatable experiments can be performed by multiple independent investigators.

Regulation and Non-Genetic Adaptation

As discussed, living organisms have the ability to pass information about their designs to descendants by means of their genetic data. Evolutionary mechanics is all about how this data is transmitted and how variants in genetically specified design are selected and propagated. However, it is very common for an evolved organism design to possess the ability to adjust itself or adapt in real-time within some range in order to better meet the requirements of current local or temporary conditions. For example, all normal humans have 639 skeletal muscles that are genetically specified. There can be genetic variation in the sizes of these muscles. You might have, by inheritance, stronger arms or weaker legs than some other individual. Our design further allows for muscle size and strength to *vary* with use. More exercise of a specific muscle causes it to become larger and stronger. Blood vessels grow to accommodate the larger muscles. These are ***acquired*** changes that ***do not*** change inheritable genetic data and therefore are not inheritable but ***do*** affect the organism's ability to cope with its external world and therefore fitness.

Intelligence is the capability to *acquire* information from the external world through *sense* mechanisms and use it to *modify* an organism's behavior in such a way as to improve survival or propagation. Immunity involves the ability of an organism's design to *sense* pathogens and *acquire* resistance to those pathogens.

Even plants can non-genetically adapt: roots grow toward water, sunflowers turn toward the sun.

All designs for non-genetic adaptation involve the following system architecture:

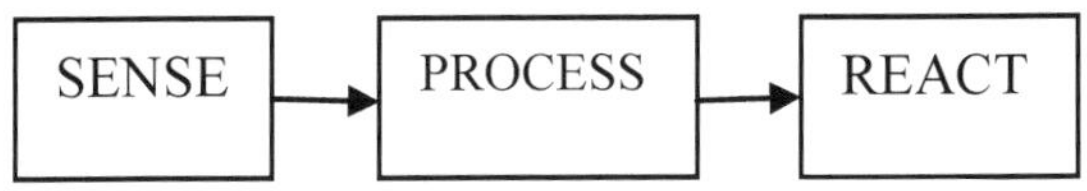

Figure 7 Non-Genetic Adaptation

All capabilities for non-genetic adaptation need to have the ability to sense a condition (generally an external condition or resulting from an external condition) for which temporary adaptation would be useful, process that information in some manner, and then effect the actual physical change necessary.

Animals are of course collections of such systems. The ability to move is of very limited value without the ability to sense external conditions, produce logical decisions from that information, and produce movement based on those decisions.

The foregoing is important in the present context for two reasons:

1. Non-genetic adaptation, especially of a long-term nature as in intelligence or immunity, represents an issue for evolutionary mechanics theory. How do acquired fitness advantages interact with genetically directed fitness characteristics? Do acquired advantages compete with genetically directed advantages in the evolution process? If not, why not?

2. If aging and lifespan limitation is indeed a necessary biological function, would we not expect an aging mechanism to have the ability to non-genetically adjust in response to temporary or local external conditions? Is there indeed evidence of such *lifespan regulation*?

These issues will be discussed further.

Mammal Life Cycle

The graph below shows the life cycle of a typical mammal and describes the animal's ability to survive under wild conditions as a function of age. As shown, this capacity increases from zero to a maximum near the age at which it is a mature adult and then declines because of the deteriorative effect of aging until eventually the animal has zero survival capacity. The animal is first able to reproduce at an age somewhat earlier than the age of physical maturity and then experiences mating seasons that are synchronized to external cues in such a way that mating seasons occur at a particular time of year.

Everybody agrees that the growth phase and the reproductive events are "programmed" by complex evolved biological mechanisms. The major issue to be discussed in this book is whether the senescence phase, including deterioration in reproductive functions, is also programmed or instead arises in a different way.

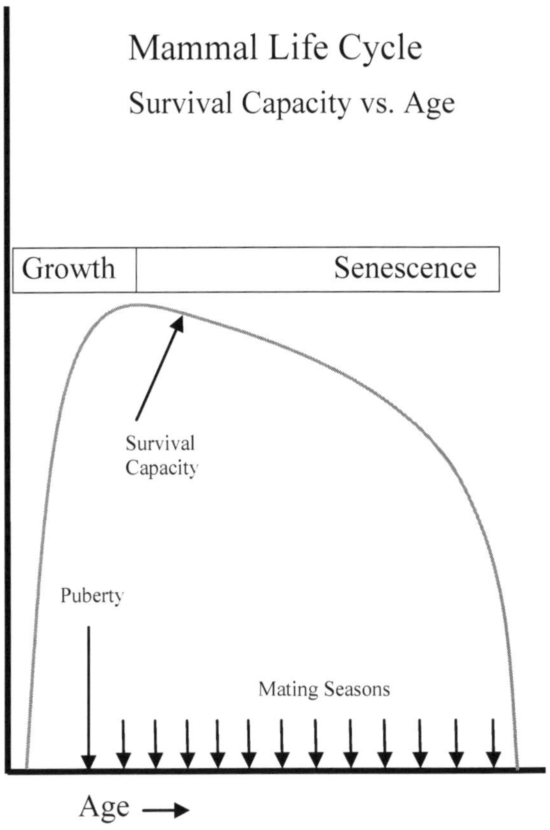

Figure 8 Mammal Life Cycle

3. Historic Theories of Aging

The following theories of aging were developed following Darwin's publication of the theory of natural selection and in response to "Darwin's dilemma." For reasons indicated, relatively few current scientists believe in the historic theories.

Weismann's Theory of Programmed Death

Figure 9 August Weismann

August Weismann (1834 – 1914) was a German biologist who published a paper[3] in 1882 suggesting that "programmed death" was a genetically programmed, evolved characteristic, (an adaptation), and that this characteristic had evolved through natural selection because it conveyed a benefit to the species even though it had a negative effect on individual fitness. Weismann's thought was that by removing older members of the population, programmed death provided more resources (such as food and habitat) for younger members. The younger animals were presumably one or more generations more evolved than older animals. Programmed death therefore shifted resources from less evolved to more evolved animals and thus improved the species' ability to evolve, that is, it improved its ability to adapt to changes in its external world through natural selection. A species having programmed death could adapt more rapidly to changes in its world and thus have an advantage over competing species or populations. Weismann's theory had a major benefit: it explained the inter-species differences in lifespan. If aging was an evolved characteristic, then we would expect the sort of major species-specific variations that we observe in other evolved characteristics.

Because of two problems, both involving incompatibilities with Darwin's mechanics, biologists of the time largely discounted Weismann's theory:

1) Is it really feasible for a species-benefiting characteristic having a negative effect on individual fitness to evolve? Critics felt that the mechanics of evolution would preclude this. Would not an individually adverse characteristic "select out" on a time scale that was very short compared to the time required for a species-benefiting characteristic to "select in"? Wouldn't an even very minor individual fitness disadvantage override an even major species advantage?

2) Is there really even a species benefit to death? Weismann's proposal that death, *per se*, was the species desirable characteristic, was attacked on the grounds that most

animals in the wild essentially never lived long enough to die of old age and that therefore the alleged benefit of programmed death could virtually never have been expressed and could not have driven evolution.

Weismann's theory violated an implicit requirement of the theory of natural selection, namely, that in order to be selected a trait has to be *expressed* in such a way that it affects survival or reproduction. Natural selection works by the differential in lifespan between animals that have a beneficial characteristic and those that do not. If almost all the animals died of other causes before the age at which "programmed death" was activated, then it could not have affected survival and thereby natural selection. Of course, Weismann's theory additionally violated the individual benefit requirement.

The bioscience available in 1882 did not uncover direct evidence of programmed death and, presumably because of intense peer pressure, Weismann eventually recanted.

Variations on Weismann's theory (to be discussed) have been recently reintroduced based on subsequent alternative evolutionary mechanics theories and new discoveries.

Fundamental Limitations - Accumulation of Damage Theories

One obvious way out of Darwin's dilemma would be to declare that aging was *not* an evolved adaptive characteristic but was instead a defect, a result of some fundamental property of life, or some fundamental physical limitation, or at least some process other than natural selection.

As years passed, many such theories appeared:

-Nuclear background radiation (or other disrupting influence) could cause cumulative, gradually increasing, damage to DNA or cell structure, or some other critical function, causing aging.

-Some poisonous byproduct of life processes could accumulate, gradually causing deterioration.

-There could be some fundamental limitation on the number of times cells can divide.

-Mechanical devices inevitably wear out eventually. Would this not also apply to living organisms?

-Some essential and irreplaceable ingredient could be gradually used up.

-The laws of entropy say that everything goes from an ordered to a less ordered state as time passes. Could not aging be an example of entropy and therefore a fundamental property of life?

Suppose we owned a fleet of automobiles. We would expect the possibility of early malfunctions in our cars (often actually referred to as "infant mortality"). Once these are corrected, we expect a period of relatively trouble-free operation. During this period, the occasional malfunction occurs. These malfunctions tend to differ greatly from case to case. One car might have radiator failure. Another might have an oil leak. The symptoms these malfunctions display also vary greatly.

However, eventually, all the cars start to develop problems. These problems tend to be more common to all the vehicles and display more common symptoms. Eventually all the cars will start to use more oil. They all start to show signs of corrosion, wear, and "aging" including deterioration of rubber, leather, paint, and polished surfaces, as well as increased frequency of mechanical failure.

Aging in the automobiles is the result of *accumulated damage.*

Wear, or the mechanical removal of material in microscopic increments, usually due to rubbing contact between parts, is one form of accumulated damage.

Oxidation, which causes deleterious accumulative changes in the physical properties of metals, rubber, leather, and other materials, is another cause.

Stress corrosion or metal fatigue, which results in parts becoming progressively weaker with use, is another cause. Microscopic changes in the molecular structure of parts occur because of the stress of normal operation. These changes cause the parts to become progressively weaker and therefore cause the *probability of failure to increase* as the machine becomes older.

The *external appearance* of our automobiles changes in a characteristic way with age. Bright surfaces tend to become dull. Colors tend to fade. Colors tend to become less uniform with spots and other variations apparent.

The *performance* of our automobiles also tends to deteriorate as they become older.

The Human Perspective on Aging

The *human experience of aging* rather precisely duplicates our automobile experience. Again, infant mortality is followed by a period in which the occasional "malfunction" occurs. Here again, these malfunctions tend to differ greatly between individuals and have different symptoms. One person might develop appendicitis; another might have a gall bladder problem. However, eventually, people develop common conditions with common symptoms and problems become more frequent.

The appearance of humans changes in a characteristic way with age. Some people can tell the age of another person within a few years by means of appearance alone. Human appearance changes are similar to those of the automobiles. Colors fade and become less uniform.

The performance of humans deteriorates with time. People become weaker and slower.

It is therefore no surprise that many people, including many physicians, think of aging as a "biological wearing out" process or otherwise the result of incrementally accumulated damage. People who are familiar with human aging but not as familiar with aging in other species generally subscribe to some variation of an accumulation of damage theory.

In our automobiles, aging is the result of the fundamental properties of the materials used in their construction. To an automotive engineer, the properties of materials are unalterable, a "given." If better materials existed that did not have any disadvantages, all automobiles (and sewing machines and bicycles) would be made of these materials. If, in our cars the water pumps tended to fail prematurely, engineers would attempt to strengthen the water pumps, perhaps by using more material in their construction. Similarly, if transmissions had a problem, a transmission-specific solution would be found. However, there is no single thing that can be done to simultaneously improve both water pumps and transmissions.

Following this sort of logic, people who believe in accumulation of damage theories for biological aging also generally believe that there is no possible *treatable* common factor in aging. The common factor between all the various manifestations of aging, accumulation of damage, is a fundamental property of life. We can find treatments for individual manifestations such as heart disease, cancer, and stroke but there is no possible treatment for the common factors in aging.

The Biology Perspective on Aging

Biologists, naturalists, and others familiar with a large number of different non-human species eventually determined that these generic *accumulation of damage* theories were not credible because of the following reasoning.

Living organisms are dramatically different from human constructs such as our automobiles. One major difference is that organisms are *self-constructed*. Living things construct themselves out of raw materials starting from a microscopic cell. If they can construct themselves, why could they not also repair themselves? Repair and maintenance certainly sounds much easier to do than the original construction. There are many obvious examples of just such self-repair. Fingernails, hair, and claws grow to replace worn items. Teeth are replaced in many animals. Damaged tissue heals (to some extent, which varies from species to species). Some creatures can replace a lost limb. Cells that are subject to rapid damage and degradation such as red blood cells and epithelial cells are replaced.

Wear or other accumulation of damage due to "normal operation" rather than the simple passage of time also does not appear to apply to animals. Exercise *strengthens* muscles and bones, generally improves health and increases lifespan. Extended inactivity *weakens* muscles and bones, generally detracts from health, and decreases lifespan. This is not the behavior one would expect if wear or other similar mechanism were responsible for aging. There are many other indications, to be described, that "normal" stress does not cause aging.

Entropy also does not represent a justification for aging in living organisms. Entropy can be reversed by the application of energy. If we construct anything, fill a tank with water, or even collect things together we are reversing entropy. Living things continuously reverse entropy by creating new structure from random components. Growth of any living thing reverses entropy.

Someone might point out that there could be some general limit on the ability of an organism to perform self-repair. Once this limit was reached, damage could accumulate. This could explain why the human experience of aging so precisely resembles our automobile experience. Some mechanisms that have been suggested as "causing" aging in humans, such as oxidation, are very similar to the mechanisms that cause aging in automobiles.

However, there does not appear to be any direct evidence that such a limitation results in the current human lifespan limit and there is substantial evidence that it does not, as described below.

The other main problem with the accumulation of damage theories is that they do not explain why aging so resembles an evolved characteristic as described in Chapter 1. Specifically, how is it that some species have dramatically different lifespans than other, similar, species? Why would the accumulation of damage in a crow be so different from that in a parrot? Why would a desert mouse wear out 100 times more rapidly than a human?

Efforts were made to find some common, simple factor that would explain the lifespan differences between different species. One such factor that appeared promising was size. Larger animals tend to live longer than smaller animals. Aquarium owners know that larger fish tend to have longer lifespans than smaller fish. Unfortunately, there were many gross exceptions. Parrots and elephants have similar lifespans.

Another possible factor was metabolism. Possibly some animals, in effect, "burned their candles" more rapidly than others. Could this explain why a mouse had a lifespan so much shorter than that of a turtle? Unfortunately, there were many exceptions here also. Crows and parrots have greatly different lifespans even though they are both flying, high-metabolism animals of similar size.

Living organisms do share many common fundamental properties of life. All organisms on Earth have extensive common chemistry and biology. However, the *differences* between different species are presumed to be the result of evolution. Although there are, no doubt, fundamental limitations on lifespan, (nothing is infinite), the differences between species suggest that observed lifespans are not the result of any fundamental limitation or any generic process that affects different species equally.

In terms of the entire biosphere, there is yet another problem. Incremental accumulation of damage could plausibly explain the sort of gradual deterioration we observe in humans and many animals, at least when we ignore the inter-species differences. However, there are species such as salmon and bamboo (see Chapter 6) in which lifespan is controlled by mechanisms that do not involve "gradual degradation" and therefore do not fit with accumulation of damage theories. A comprehensive theory cannot ignore these species.

Reliability theories of aging represent a variation on wearing out theories and have some current popularity[4]. These theories hold that aging is the result of the accumulation of microscopic defects in a manner similar to the aging process in automobiles or other machines. These theories ignore self-repair and fail to explain why some very similar organisms have very different lifespans, why organisms with very similar microscopic design (e.g. mouse and man) have grossly different lifespans, why stress increases lifespan, and why some species have lifespan regulation mechanisms that are not gradual and accumulative.

Stochastic theories of aging similarly hold that aging is simply the result of random and accumulative changes.

Although most biologists no longer believe in the simple accumulation of damage theories, a majority of Americans and many physicians do believe in one of these theories. This is highly significant in connection with their attitudes regarding anti-aging research as described in Chapter 7.

4. Modern Non-Programmed Theories of Aging

Most present-day biologists, and many medical researchers, physicians, and health scientists believe in one or more of the following non-programmed theories of aging. If you take a college level biology course, you may be taught that one or more of these theories is "generally accepted."

Evolutionary theories of aging are those based on evolutionary mechanics theory. As described earlier, it was obvious to biologists and others familiar with many species that lifespan was extremely unique to individual species. Efforts to find some generic, species-independent explanation for aging had failed. Clearly, lifespan had to either be a part of the design of each individual species or somehow very closely related to and dependent on the specific design of the species. Since theories of evolutionary mechanics specify what sorts of design characteristics can or cannot evolve, and how organism designs result from an evolutionary process, it was reasonable and necessary to invoke evolutionary mechanics theory in searching for an explanation.

Let us review the circumstances surrounding aging theory in 1950:

Weismann's theory had failed because of major undefended incompatibility with Darwin's mechanics. By 1950 both Darwin's theory of species origin and the orthodox theory of evolutionary mechanics had nearly universal acceptance in the scientific community. It was apparent that many, possibly most, biologists would summarily reject "out of hand" any theory of aging that conflicted with Darwin's mechanics. This situation continues to a lesser extent today, (see Chapter 7), and some theorists continue to reject non-compliant theories based entirely on evolutionary mechanics considerations.

The generic accumulation of damage theories had failed because of incompatibility with observed characteristics, notably the inter-species lifespan variations.

Therefore, in 1950, the scientific situation was that the fundamental nature of aging was a total mystery, "an unsolved problem of biology."

The popular situation was much less confused. Anybody with "half a brain" could see that people wore out in virtually exactly the same way as the family Ford or Aunt Hattie's sewing machine.

Because of these circumstances, it is fair to say that new scientific theories of mammal aging needed to meet the following criteria, in order of decreasing importance:

1. Maintain compatibility with traditional evolutionary mechanics. As a minimum, plausibly claim at least some compatibility with orthodox theory.
2. Explain the inter-species lifespan variations, at least in mammals. Compatibility with other organisms was less important. Obscure, bizarre, lifespan observations could be ignored. Specifically, ignore instances of biological suicide.

All the modern non-programmed theories meet these criteria.

These theories are often called "evolutionary," "non-adaptive," theories of aging because they, in effect, combine natural selection with "accumulation of damage." Natural selection explains why animals live long enough to reproduce and "accumulation of damage" explains why they age after reproducing when aging apparently has less effect on fitness.

A major difference between the accumulation of damage theories and the newer theories is that in the non-programmed theories, the factors that cause aging are genetically transmitted but not "genetically programmed." The term "genetically programmed" is used to mean an "adaptive" function such as Weismann's theory. Because aging traits are genetically transmitted, inheritance of aging traits can be explained by the traditional theories.

Medawar's Concept: Force of Evolution Declines with Age

Sir Peter Medawar (1915 – 1987) was a noted British professor of zoology and anatomy at the University of London who won the Nobel Prize in medicine (1960) for his work on acquired immunological tolerances. In Medawar's 1952 paper[5], *An Unsolved Problem of Biology,* (originally presented as a lecture at University College London) he presented an ingenious theory, which, in effect, combines the properties of the accumulation of damage theories with Weismann's evolved characteristic theory. In effect, Medawar suggested that survival of the fittest only applied to young organisms and that the evolution process was not affected by the fates of old organisms where "old" was defined relative to age of initial reproductive capability.

Medawar suggested that the force of natural selection decreases once an organism reaches an age where it has had some opportunity to reproduce. If, for example, some trait of an animal tended to be fatal prior to puberty, that trait would be very strongly selected against because most animals having that trait would die before having any progeny and would therefore not pass their adverse trait to descendants.

If, on the other hand, an animal had a trait which caused a fatal effect only after the animal had reached sexual maturity, survived to least one mating season, mated, had progeny, and nurtured and protected those progeny long enough for them to become self-sufficient (assuming it is an animal that nurtures and protects young), the effect of that trait on fitness would be relatively insignificant. The negative trait would only affect the animal's ability to survive yet longer and have subsequent descendants. Such a trait would apparently only weakly affect fitness. Aging seemed to fit this description.

Figure 10 Peter Medawar

Medawar proposed that even if an animal did not age, that is, did not have an increased probability of death as a function of age, the numbers of adult animals of any given age would decrease exponentially because of deaths due to other causes such as predators, environmental conditions, etc. Some constant percentage of the animals of any given age would be killed in any given time period. As a metaphor, he used the random breaking and replacing of test tubes. If, in a lab with hundreds of test tubes, when a test tube is broken it is replaced with a new (age = 0) test tube, then after a while the number of test tubes of a given age in the test tube population will decline exponentially with age. The graph below shows the kind of survivor's curve (number as a function of age) that would be expected for the test tubes (or ageless animals) as presented in Medawar's paper. This behavior is known as an *exponential decline* and is characterized as having a *half-life*. If half of the animals die in four years (as shown here), then half of the remaining half will die in another four years, and so forth. Very old animals would be very rare even in a non-aging species.

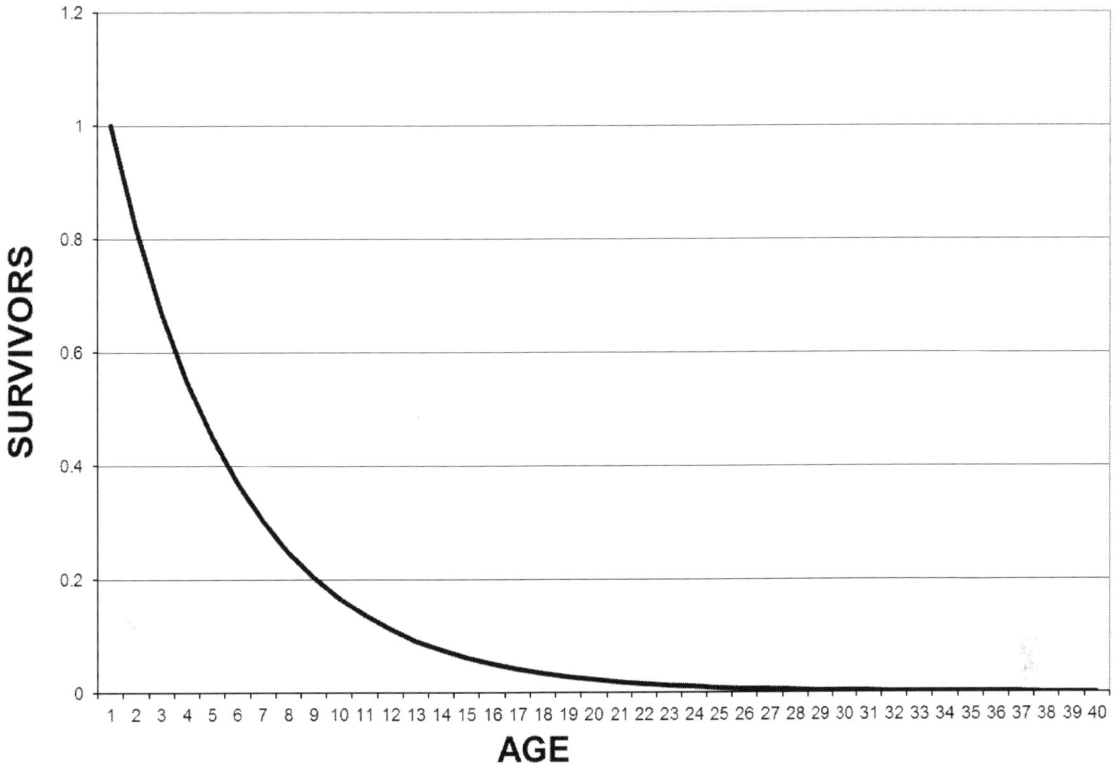

Figure 11 Medawar's Model - Survivors vs. Age

The number of progeny any given non-aging animal would produce was assumed by Medawar to increase linearly with age. An ageless animal would tend statistically to have the same number of progeny every year from puberty to death. The total number of progeny (on average) produced by animals, as a function of their age, would therefore look as shown below in which total descendants produced by an animal are zero until puberty (defined here as age of initial reproductive capability of any organism) and then increase with age. Puberty is shown here as occurring at age 6. It is assumed that reproductive vigor in an ageless animal would not change with age. We could have assumed that general reproductive effectiveness and aging were different, independent phenomena. However, an evolved characteristic that causes a decline in reproductive effectiveness has the same problems with Darwin's mechanics as aging and the same arguments, pro and con, apply.

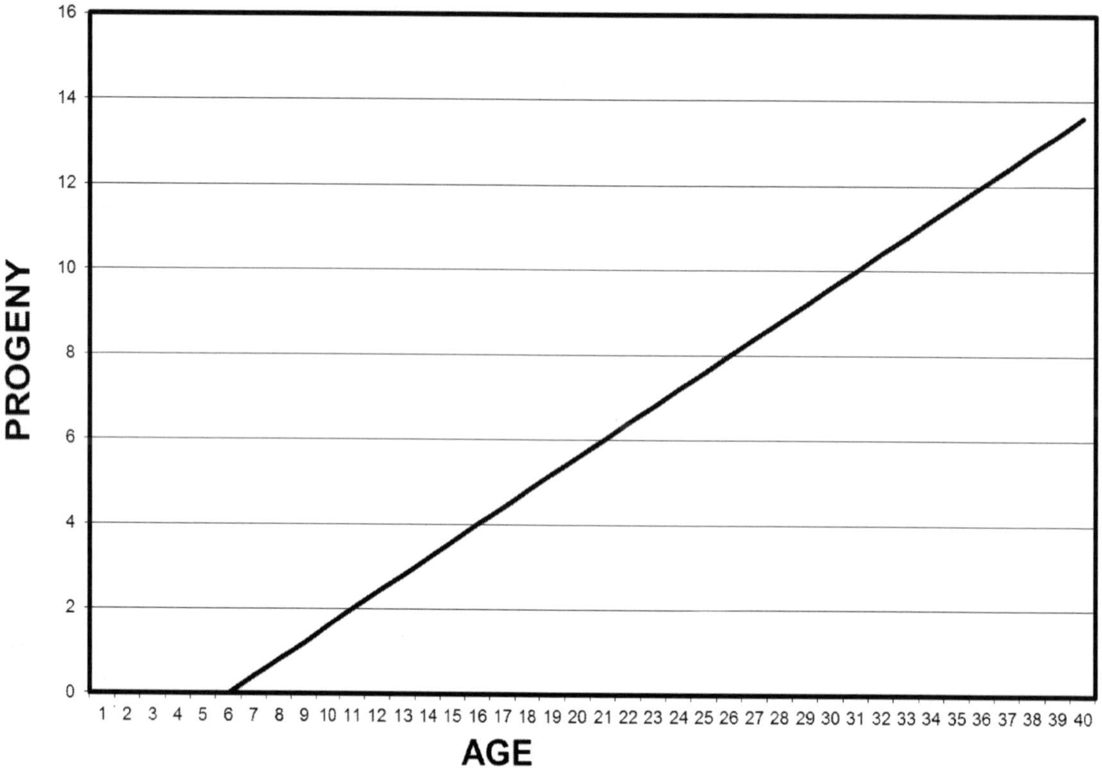

Figure 12 Medawar's Model - Progeny vs. Age

If we multiply the number of progeny produced by animals of a given age by the number of animals at that age we can determine the *reproductive effect* contributed by each age group or *cohort*. Since the number of animals is exponentially decreasing with age and the number of progeny is only linearly increasing with age, the reproductive effect of older animals (and therefore their apparent evolutionary impact) declines. Graphically, this would look like the figure below in which reproductive effect rises from zero at puberty to a maximum and then declines.

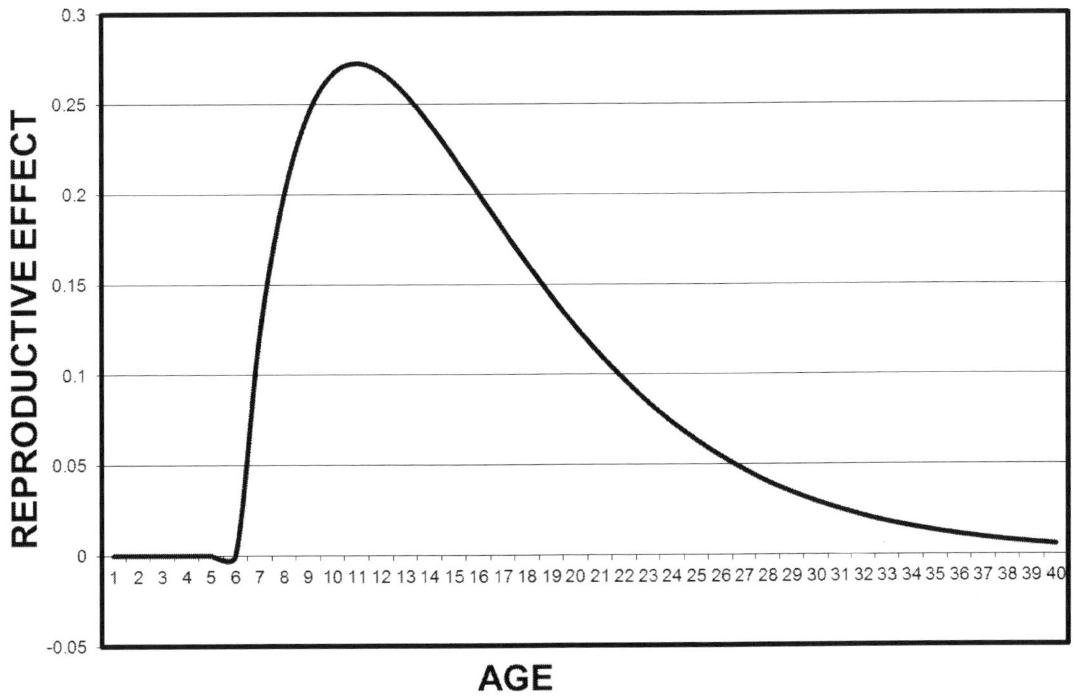

Figure 13 Medawar's Model - Reproductive Effect

These curves represent Medawar's model of a non-aging species. Medawar's paper was almost entirely devoted to the development of this model. This proposal was a somewhat brave and desperate proposal on Medawar's part. It was brave because Darwin's survival-of-the-fittest concept fit perhaps 99 percent of all biological observations that had ever been made and Darwin was probably the most respected single individual in biology. He could have been jeered out of the room. It was desperate because in more than 90 years, a plausible Darwin-compatible explanation for the multi-species lifespan observations had not appeared and it was apparent that a major change in approach was needed. Medawar's paper was titled "An unsolved problem of biology."

The shapes of these curves would tend to vary with species. Some species, subject to high death rates due to predators or other cause would need a correspondingly high birth rate to survive. Other species would need a lower birth rate.

Medawar proposed that aging was caused by random mutations causing adverse aging characteristics. In effect, aging was caused by an assortment of genetic diseases, each of which has adverse symptoms only at advanced ages. Medawar discussed in this connection human genetic diseases such as Huntington's chorea, which does have increasing symptoms with age. While Huntington's and other genetic diseases each individually only affect part of the population, it is easy to imagine that other adverse genetic conditions could have spread to encompass essentially the entire population and thus be considered a "natural" part of aging.

Medawar's "mutation accumulation theory" was written in response to, and as an alternative for, Weismann's earlier programmed death theory, and neatly reverses the main disadvantages of Weismann's theory turning them into advantages. Most importantly,

Medawar's theory does not require a violation of Darwin's individual benefit requirement. By 1952 (and still to a lesser extent today) many biologists considered the natural selection theory to be essentially infallible, "a given." Second, if you reject Weismann because most wild animals do not live long enough to die of old age, and that therefore programmed death cannot be an evolved characteristic, then you should accept Medawar's idea that mutations can accumulate causing death of old age. These two ideas are opposite sides of the same coin.

It is important to note that Medawar considered that, for any given species, there was essentially zero evolutionary motivation for that species to have a longer lifespan. For virtually any species, we can find another similar species that has a shorter or longer lifespan. Therefore it was clearly possible for any given species to evolve a lifespan that precisely matched its need for lifespan. If conditions changed in such a way that the species needed a longer lifespan, the species would evolve a longer lifespan. If they changed such that the species did not need to live as long, its lifespan would decline. Darwin's tiny steps idea says that the evolution process can respond to a very small benefit, so we could use the phrase "effectively zero" to describe the evolutionary benefit of living and reproducing longer than a species-specific lifespan. An even very slight benefit would cause lifespan to increase.

By combining evolution (to explain early longevity) and accumulation of damaging mutations (to explain lack of later longevity), Medawar neatly sidestepped Darwin's dilemma, while still explaining some of the inter-species differences in aging. Since Medawar's theory tied aging to sexual maturity and reproduction, it provided a much better fit to observed characteristics of animals.

Medawar's theory is still respected today.

His concept encompasses two ideas that are very widely incorporated in subsequent programmed and non-programmed theories of aging:

First, that there is a relationship between the age at which an organism is first capable of completing a reproductive activity and the lifespan needed by that organism. We can all agree that an organism that died of old age prior to reaching puberty would not be logically viable.

Second, that there is a relationship between internal causes of mortality and fitness degradation (aging) and external causes of mortality and degradation (diseases, predators, harsh environment, lack of food, etc.). We can all agree that there would be no evolutionary advantage to overcoming internal limitations to living beyond the age at which 100 percent of the individuals in a population could be expected to be dead from external causes of mortality.

The major difficulty with Medawar's concept is that internally caused degradation and death generally occurred far too early than appeared to be justifiable by external limitations.

It is important to note that, unlike the earlier wear out or entropy theories, the mutation accumulation theory does not suppose any fundamental, inescapable, cause of aging. Aging is the result of adverse mutations. If these mutations could be removed or contravened, longevity could be extended, perhaps indefinitely.

Furthermore, aging affects only what might be described as "maintenance" functions, namely, those activities needed merely to maintain the condition of an already developed, fully functional, adult, organism. The scope and difficulty of maintenance would appear to be relatively minor, even trivial when compared to the activities involved in the growth, development, and normal day-to-day functioning of an organism.

Many of the activities involved in maintenance, such as cell division and replacement, would appear to largely duplicate those involved in the original growth and development. Therefore, it is reasonable to believe that a relatively small number of genes are exclusively associated with the maintenance function, such as genes that control initiation of cell division only in a maintenance context. It is only these genes that are affected by the adverse mutations.

Finally, in relatively longer-lived animals, only those few maintenance functions involved in relatively long-term maintenance are adversely mutated since the shorter-term functions are already fully operational. Even the shortest-lived mammal would have needs for maintenance. Wounds heal. Hair and claws grow. Short-lived cells are replaced.

Medawar's theory therefore suggested that medical intervention that contravened the relatively small number of adversely mutated functions was at least a possibility.

We know from modern genetics (See Genetics.) that related organisms such as mammals have a very high degree of similarity in their genes, which leads to the conclusion that they share very similar logical processes or "genetic programs." The differences between different mammals are directed by relatively minor genetic differences that in turn cause differences mainly in degree or magnitude rather than in the logic. In other words, mice and men probably have the same "maintenance program." It seemed plausible that the program in mice is less aggressive and effective than the program in humans so lab mice live perhaps 14 months after reaching maturity and humans live about 60 years after maturity.

The idea that mutations could occur that would cause adverse effects has been verified by substantial work that has been done in an effort to understand human genetic diseases (see Genetics). Many human diseases have been traced to errors that have occurred in genetic code. Symptoms of some genetic diseases are even age-related and tend to increase with age.

There are problems with the idea that such mutations cause aging as put forth in the mutation accumulation theory:

The mutation accumulation theory only works if the fitness effect of aging is effectively zero. Mutation accumulation says that "absence-of-aging" does not evolve because the beneficial effect of absence-of-aging is small enough that mutations that contravene absence-of-aging can accumulate and not be selected out. This would *also* apply to *any other trait* that had the same or less beneficial effect. If "slightly longer claws" is very mildly beneficial then "slightly shorter claws" is only very mildly adverse. Evolution is still able to evolve slightly longer claws. Darwin's theory of tiny incremental steps requires that very small beneficial characteristics can evolve. Medawar's hypothesis depends on the idea that aging can exist solely because of the declining fitness effect of adverse events with age and therefore depends on the idea that aging has a negligible fitness impact.

Although "death of old age" probably only occurs frequently in species that have few predators, aging in mammals obviously has effects other than death that would affect fitness and therefore death rate. Aging in mammals affects strength, speed, agility, and other factors that affect fitness even in relatively young animals. It therefore does not appear plausible that aging has a negligible effect on fitness.

If we assume a Medawar scenario in which a species (e.g. mouse) lives in a brutally vicious world where virtually no animals survive long enough to die of old age, then even a very minor difference in a survival trait (speed, strength, etc.) would presumably influence the probability of survival. Such minor differences due to aging could plausibly be expected to appear at very young ages.

Medawar's own model, the broken test tubes, does not result in zero beneficial effect from living longer but, as shown in the reproductive effect sketch above, shows that the benefit of living longer only *approaches* zero as age approaches infinity.

If we assume a species (e.g. elephant) that has relatively few predators and therefore lives a relatively peaceful existence in the wild, then presumably death of old age is a fitness factor.

These issues led to the subsequent development of the antagonistic pleiotropy theory described in the next section.

Another problem is that a number of diverse non-mammal organisms (salmon, octopus, and bamboo – see chapter 6) display instances of death closely following an act of sexual reproduction. Death in these species appears to be *controlled* by the reproductive function or controlled by whatever triggers reproduction as opposed to calendar age. Aging in mammals, as a gradual, diffuse, and multi-tissue degradation loosely tied to sexual maturity, could plausibly result from random mutations variably degrading a family of beneficial maintenance characteristics. However, this scenario does not appear to work for bamboo and salmon, which exhibit what appears to be programmed death tied directly to reproduction and clearly not associated with a generalized maintenance function. How could suicidal behavior result from the random mutation degradation of a beneficial characteristic? What beneficial characteristic was degraded to result in biological suicide?

The basic problem is that mutation accumulation is too simple a mechanism to explain the detail in the observed aging processes of different species. As we will see in later sections of this book, the traditional model of non-aging species also grossly understates the negative fitness effects of aging in actual animals.

Medawar's idea is actually more *population-oriented* and less individual-oriented than Darwin's individual-benefit-only theory despite claiming compatibility with orthodox Darwinism. According to Medawar and his test tube model, an older immortal mature animal would be functionally identical to its younger mature self. It would have identical fitness and ability to survive and reproduce. The difference between younger and older animals is that the *populations* of older animals would be smaller and that the *smaller populations* would have less effect on the evolution process.

Some gerontologists think of Medawar as a major prophet, essentially the father of modern gerontology. Indeed the more recent programmed aging theories are based on the idea that the evolutionary value of survival and reproduction declines with age.

Williams' Antagonistic Pleiotropy Theory

George Williams, then a professor at the Michigan State University, published a paper[6] in 1957 titled *Pleiotropy, Natural Selection, and the Evolution of Senescence*. *Pleiotropy* is defined as a situation in which a single *allele* or form of a gene (see Genetics) may affect more than one trait. In human genetic diseases, a defect in a single gene typically can affect a number of traits and have simultaneous diverse symptoms such as nerve and vision problems, bone deficiencies, and skin changes. In general, a single gene can be activated in more than one tissue and therefore a defect in a gene can affect more than one tissue.

Williams specifically criticized Medawar's assumption that the fitness effect of aging was negligible:

> "No one would consider a man in his thirties senile, yet, according to athletic records and life tables, senescence is rampant during this decade. Surely this part of the human life-cycle concerns natural selection. … It is inconceivable in modern evolutionary theory that senescence, such as operates in man between the ages of thirty and forty is selectively irrelevant."

Williams proposed that aging was caused by the combined effect of many pleiotropic genes that each had a beneficial effect in an animal's youth but had an adverse side effect in older age.

Williams' concept was similar to Medawar's in that it built on the idea that adverse effects have a progressively smaller impact on fitness as an animal gets older. A gene resulting from natural selection could have a rather catastrophic negative effect on an older animal if the negative effect was balanced by an even relatively minor beneficial effect on younger animals.

Williams' theory, like Medawar's, provided a better fit to the observed inter-species variations in aging than the accumulation of damage theories while simultaneously avoiding Darwin's dilemma and did not depend on accumulation of adverse mutations in equilibrium with out-selection. Williams' theory avoided the apparent difficulty in the mutation accumulation theory that required the negative fitness effect of aging to be negligible because the assumed beneficial effect of the pleiotropic genes balanced the negative (aging) effect. Williams predicted that species with younger age of sexual maturity and more vigorous reproduction traits would tend to have shorter lifespans.

One consequence of Williams' theory was that prospects for any significant treatment of the fundamental causes of aging were considered negligible because of the assumed large number of antagonistic genes and the assumption that the harmful aging genes had beneficial and probably essential functions in youth. Hopes that only a relatively small number of factors causing aging would be eventually found and successfully treated as suggested by the mutation accumulation theory were therefore, according to Williams, misplaced. Williams said it this way:

> "Any such small number of primary physiological factors is a logical *impossibility* if the assumptions made in the present study are valid. This conclusion banishes the "*fountain of youth*" to the limbo of *scientific impossibilities* where other human aspirations, like the perpetual motion machine and Laplace's 'superman' have already been placed by other theoretical considerations." [*Emphasis* added.]

Clearly, believers in Williams' theory, which is still popular with some people, also believe that anti-aging research is a fundamentally foolish endeavor, a chase after the fountain of youth.

Williams' idea is a recognition of the idea that the ***net*** negative evolutionary effect of aging, counting all tradeoffs and all aspects of aging such as deterioration and decline in reproductive ability, would need to be effectively zero. If it was not, presumably the organism would evolve a longer lifespan. The reason for the "effectively" qualifier is that it is widely recognized that evolution and retention of a trait requires that the trait produce and continue to produce at least some minute evolutionary benefit. However, in accordance with Darwin's and subsequent thinking, this benefit could be tiny.

Over the years many theories were developed suggesting that aging was rigidly linked to some beneficial property. Aging could be a rigidly-linked side-effect of some property that fights cancer. Aging could be a rigidly-linked side-effect of growth. Aging could be a rigidly-linked site-effect of [insert your own favorite beneficial property here].

The antagonistic pleiotropy theory (along with other rigid-linkage theories) has a number of difficulties.

One problem with the antagonistic pleiotropy theory is that, according to Williams, the force of natural selection, although apparently progressively smaller in older individuals is not zero, even according to Medawar's model. This was the genesis of the antagonistic pleiotropy theory. Evolution would therefore presumably be trying to find ways of accomplishing the beneficial effects *without* the adverse (aging) side effects. Why would it not succeed? One obvious answer is that *there is no way* to accomplish the presumably essential beneficial effect without the side effect, (a conclusion that is very pessimistic regarding successful treatment of aging and age-related diseases). This conclusion requires that we believe that the pleiotropic relationships are *indefinitely rigid*, that is, that they have withstood the force of evolution for at least the time that sexually reproducing organisms have existed on Earth. This is because Williams offers no suggestion to the effect that the value of extended life would be less in the more primitive organisms that constitute ancestors of mammals. Evolution would have been trying to overcome the pleiotropic linkage in these ancestors and in their ancestors for billions of years. Had it been successful, mammals would have inherited the advantage without the disadvantage.

Williams' pleiotropy argument is based on a particular aspect of modern genetics science. *Other* aspects of genetics science provide extensive arguments contravening his conclusion that pleiotropy causes *indefinitely rigid linkage* between individually adverse and individually beneficial design characteristics (to be discussed). It turns out that pleiotropy is only one of several genetic characteristics that lead to linkage.

Additionally, it seems extremely implausible that this "unavoidable side effect" difficulty would only affect maintenance of the condition of an adult organism when the tasks that have to be performed in the development and growth of an organism are apparently so much more complex and difficult. Further, even considering only mammals, random rigid linkages do not appear to have prevented different mammal species from evolving grossly different designs that adapt them to their individual circumstances.

Another difficulty with the antagonistic pleiotropy theory is the very great variation in observed aging characteristics between otherwise similar animals such as similar birds and fish. If aging is a side effect of genes that have a beneficial effect in youth, then why would the adverse side effects of such presumably similar genes be so different? If the side effect is an unavoidable consequence of the beneficial effect, then why would a very similar animal need a different gene and display a different side effect?

The antagonistic pleiotropy theory also appears to have a fundamental conflict with modern genetics (see Genetics) as follows: The "antagonistic" aspect of the theory works. Essentially any feature of an organism has tradeoffs. Faster is antagonistic to stronger.

"Pleiotropy" works. Genes are known to affect multiple tissues. The problem is that this theory proposes a sort of *time-sequential* antagonistic pleiotropy in which the *same* gene has beneficial net value at one stage in an animal's life and adverse net consequences at another stage.

The difficulty with this is that we now know that genes are activated and deactivated in accordance with a very complex genetically controlled system of logic or *program*. Essentially any gene would have adverse consequences if activated in the wrong tissue, or under the wrong circumstances, or at the wrong stage in an organism's life just as failure to activate the gene at the proper times in the proper tissues has adverse consequences. Cancer is an illustration of one of the consequences of activating genes at the wrong place or time. Many genetic diseases result from failure to activate a gene at the proper time.

The genetic program is apparently capable of coordinating the vastly different activities that must take place in the various developmental stages of an organism. Grossly different combinations of genes must be activated during, say, embryonic development than in late childhood because of the grossly different growth activities that must be performed. There does not appear to be any plausible reason to believe that the activation program would fail to deactivate a gene that would cause a problem in later adulthood if it worked so well in programming the differences between all the other stages in an animal's life, especially when the differences between other stages are so much greater than between "adult" and "older adult." Why wouldn't time-sequential antagonistic pleiotropy generally be more of a problem in early development where there is a greater need for age sequential changes?

In many ways, Williams' theory is a restatement of Darwin's circular "explanation": There must be some individual benefit compensating for the individual disadvantage of aging merely and entirely because orthodox theory says so.

Despite the multitudinous major difficulties, the antagonistic pleiotropy theory is respected among some current scientists. This has a serious negative impact on anti-aging research since, according to Williams, anti-aging medicine is impossible.

Williams' proposed several ideas that are crucial components of modern non-programmed *and* programmed aging theories:

- The net evolutionary disadvantage of aging must be effectively zero.
- Aging must somehow create an evolutionary advantage that compensates for the declining evolutionary disadvantage of not living and reproducing longer.
- Inter-trait linkages can be produced by the genomic design of organisms that make it difficult for the evolution process to remove a trait without affecting other traits.

Disposable Soma Theory

In 1977, a statistician named Thomas Kirkwood (now a biologist and professor of medicine at the University of Newcastle) published his *disposable soma theory[7]* of aging. Kirkwood's idea was that organisms only have a limited amount of energy that has to be divided between reproductive activities and the maintenance of the non-reproductive aspects of the organism (soma). Aging is the result of natural degrading processes that result in accumulation of damage but the damage can be repaired by the organism at the expense of reproductive effort. Because of the declining evolutionary impact of adverse events on older animals (Medawar's hypothesis), a tradeoff exists in which it does not make sense for an organism to invest effort in maintenance (at the expense of reproductive activity) to result in living much beyond the initial breeding years.

This theory also, in effect, combines the apparent declining force of natural selection after breeding age is reached with accumulation of damage, and additionally explains a relationship between reproduction and lifespan while avoiding conflict with the individual benefit requirement.

The disposable soma theory seems to make sense on a philosophical level but has many severe problems if one looks at it in any detail. One obvious issue is that it has not been demonstrated that "maintenance and repair" actually takes a significant amount of effort or resources when compared to the energy and resources required by the day-to-day existence of an organism much less the initial growth of the organism. Why would replacing a few cells take more effort or resources than producing trillions of cells in the first place or providing for their routine operation?

Another problem with the disposable soma theory is that it is not obvious why effort or energy spent in reproduction in an animal's early years would necessarily decrease the energy available in later years for "repair." One would think a post-menopausal woman would cease aging or even become stronger as a result of the absence of the resource drain caused by reproductive effort. Instead, damage appears to increase exponentially.

Another obvious problem is that by essentially any method of accounting females use more resources and energy in reproduction than males. Why don't males therefore live longer than females? A fan of the disposable soma theory once suggested to me that males, by amazing coincidence, spend as much extra energy and resources protecting their mates and young as females spend in reproduction and therefore have approximately the same lifespan. If you believe this, I have some stock in the Brooklyn Bridge I would like to sell to you!

Maintenance and repair appears to be largely a short-term activity. Processes such as hair and fur growth, nail and claw growth, replacement of red blood cells and epithelial cells, and wound healing are all relatively short-term and occur in time periods on the order of weeks. Needs for sleep, largely seen as a maintenance and repair function, are very short-term. Further, the resources and energy required for maintenance and repair are clearly largely associated with the short-term processes. We can readily imagine that an animal, during its lifetime, could produce a total quantity of fur, blood cells, and epithelial cells that exceeded its body mass. Even if we imagine that there are long-term maintenance processes (e.g. replacement of long-lived cells) it is obvious that the total lifetime quantity of resources associated with such a process would be trivial by comparison. The consequence is that it is extremely difficult to see how there could be an individual benefit tradeoff between maintenance and repair in later life and reproduction in earlier life.

A fundamental difficulty is that animals obtain the resources and energy they need from eating. A pregnant animal needs more resources and therefore eats more. An animal that is growing needs more resources and therefore eats more. If indeed an animal needed significant resources to perform maintenance and repair functions why would it not simply eat more to provide those resources?

Note that the idea that the *population* might be better off by expending resources on younger members as opposed to older members is incompatible with traditional evolutionary mechanics and essentially a restatement of Weismann's idea. Kirkwood was certainly *not* looking for or proposing a non-traditional explanation or agreeing with Weismann.

One way out of this problem is to assume that some unknown property that increases reproductive effectiveness somehow, for unknown reasons, decreases lifespan by causing deterioration in later years. This is very similar to the antagonistic pleiotropy theory. Aging must be a hidden unavoidable side effect of reproduction. Of course this amounts to a restatement of Darwin's "explanation:" There must be some hidden individual benefit to offset the individual disadvantage of aging. In fact, some traditional biologists such as Leonid Gavrilov, of the Center on Aging at the University of Chicago, consider the disposable soma theory to be a "version of" the antagonistic pleiotropy theory and a "widowed concept."

Common Problems with Traditional Evolutionary Aging Theories

All of the evolutionary non-programmed aging theories consider that the *net* evolutionary effect of aging is effectively zero beyond some species-specific age and that if that were not so, the species would have evolved a longer lifespan or devolved to a shorter lifespan until it was so. In the case of Williams' and Kirkwood's and many other similar theories, the *net* effect of a trait is calculated by combining the negative effect of aging with the positive effect of some linked benefit. In Williams' case we are combining the negative effect of aging (on survival and reproduction) with some rigidly linked positive effect of the same mutational change to result in "effectively zero" net result.

All of the theories based on individual-benefit-only evolutionary mechanics depend on the idea that the evolutionary importance of individuals in a non-aging population declines with calendar age and that therefore natural selection allows the existence of progressively more negative traits with age, up to and including "death of old age." This idea, the traditional model, is based on a very important assumption, namely, that the evolutionary contribution of each individual animal can be accurately described by "reproductive effect."

You will recall that the reproductive effect graph reflects the combined effect of the reproductive contribution we could expect from individual animals (progeny graph) and the rate at which we could expect the non-aging animals to die off (mortality graph). Because the population of older animals declines more rapidly with calendar age than the reproductive contribution of older individuals increases with calendar age, the net effect is that evolutionary importance declines with calendar age beyond some point that varies with age of puberty. This traditional model of a non-aging species is over-simplified and ignores a number of important characteristics of actual animals, all of which tend to increase the evolutionary importance of *older* animals.

First, the characteristics of Medawar's test tubes (and non-aging animals) were presumed to be constant and did not change with the age of the test tube. However, the characteristics of actual non-aging animals would change greatly with calendar age, at least between puberty and "maturity."

Suppose we had a group of prehistoric 15-year-old humans and another group of 20-year-olds. According to the reproductive effect concept, these two populations are equivalent and have the same evolutionary importance. In actuality, the 20-year-olds are superior with regard to essentially any survival characteristic. They are stronger, faster, and smarter. They would win in any competitive situation. It is obvious that a 20-year-old has a greater chance for survival than a 15-year-old, everything else being equal. Therefore, death rates would nominally tend to decline with calendar age in the interval between 15 and 20 as opposed to remaining constant as proposed by the traditional model. In actual animals, still additional complexity such as protection-of-young might affect this.

There is a more profound difficulty. The reproductive effect curve assumes that ten young individuals each producing one descendent have the same evolutionary importance as one older individual producing ten descendants. Because they are more mature, the 20-year-olds more fully exhibit *adult* survival characteristics. A case could therefore be made that a single 20-year-old has more evolutionary importance than any number of 15-year-olds. The reproductive effect concept does not take into account this "maturity factor."

This is the same argument made against Weismann's theory. Natural selection cannot operate relative to a characteristic that is not expressed. Adult characteristics are not fully expressed in juveniles. Therefore, adults are required in order to evolve adult characteristics. Evolutionary importance is not the same as reproductive effect.

Second, the individuals in the population of Medawar's test tubes were presumed to be identical and have identical characteristics. The characteristics of actual animals *vary*. More specifically their characteristics regarding capacity for survival vary. Therefore, their probability of survival varies.

Medawar's model assumes that the probability of death for an ageless animal is a constant, independent of calendar age. That is, older animals are just as likely to die in any given time period as younger animals. While true for test tubes, this idea is incompatible with the theory of natural selection. According to Darwin, animals that are more fit are less likely to die than animals that are less fit. Therefore, in any given time period, more of the less fit animals in a population would die. At the end of the period, the surviving animals would therefore, on average, be *more fit* than at the beginning of the period. In other words, average fitness of a non-aging population cohort *increases* with calendar age. Because older animals are more fit, they are also less likely to die relative to younger animals. The probability of death therefore *decreases* with calendar age in a non-aging population. The reproductive effect graph therefore *does not* accurately represent the relative prevalence of older individuals in a population of actual non-aging animals.

Third, more complex animals have other characteristics that tend to increase the evolutionary importance of older animals. Mating rituals and other societal behavior traits such as those that result in "pecking order" tend to reduce the relative chance that younger animals will reproduce and in some cases also reduce the relative probability of death for older individuals. The old king is far less likely to die in the war than the young foot soldier, especially if the old king does not age.

Animals learn from experience. Older non-aging animals are therefore less likely to die than genetically identical younger animals having less experience.

Finally, the increase in fitness with calendar age has a multiplying effect. Not only did the older individuals survive longer and have more descendants because they are more fit, but their descendants are also likely to be more fit and therefore survive longer, and their descendants descendants are likely to survive longer. This multiplication factor would appear to cause the *evolutionary importance of individuals to increase exponentially with age*.

If we attempted to modify the reproductive effect graph to develop an "evolutionary importance" curve based on these factors would it still decline with calendar age? Would the increasing evolutionary importance of older individuals compensate for their decreasing numbers?

It should be clear from the above discussion that Medawar's model grossly underestimates the evolutionary impact of older non-aging individuals, especially in more complex organisms such as mammals. Further, the actual shape of an evolutionary importance curve is dependent on complex interactions that vary from species to species and probably even vary depending on the species' situation regarding predators and other factors that affect lifespan. In effect, the traditional evolutionary theories of aging embrace the convenient aspects of Darwin's mechanics while ignoring the inconvenient aspects.

Could it be that nature ***needs*** aging or other lifespan control mechanism to prevent a relatively few older individuals from dominating the evolutionary process? Does aging, per se, contribute an evolutionary benefit?

These concepts are further developed in the section on evolvability theory.

Experimental attempts to confirm the traditional theories have been generally unsuccessful. For example, investigators have been unable to find a rigid, fixed connection between reproduction and lifespan as would be expected by the disposable soma theory.

So where do the non-programmed theories leave us? What collective wisdom does biology have for the rest of the world regarding the main questions listed in Chapter 1? We can summarize this as follows:

-Regarding the nature of aging: We don't really know. There are a number of different theories that attack each other. The theories all have logical flaws. Efforts at experimental confirmation have been generally unsuccessful. Aging is *still* "an unsolved problem of biology."

-Regarding the practical aspects: The general consensus seems to be that treatable common factors are either very unlikely or possibly even "impossible." Of course, the credibility of this finding is degraded by the inability to definitively determine even the basic nature of aging.

Evolutionary Value of Life Controversy

All of the aging theories discussed in this book are ultimately inherently based on one of four concepts regarding the evolutionary value of life. As discussed earlier, the "evolutionary" theories of aging assume that the value of survival and reproduction changes with age and that therefore the force of evolution directed at maintaining life varies with age, especially relative to the age at which the organism becomes capable of reproduction.

The sketch below illustrates the four different scientific concepts regarding the evolutionary value of life as a function of age relative to age of reproductive maturity. The benefit or cost of life is a measure of evolutionary force toward adapting changes in the design of an organism.

All four concepts agree that it is beneficial for an organism to live long enough to reach reproductive maturity and that degradation due to aging prior to that point would represent an evolutionary disadvantage. Further, as illustrated, lifespan beyond the minimum required for reproduction would be useful for organisms (e.g. mammals and birds) that need additional time to protect, nurture, or train their descendants. Other characteristics of specific species could affect details of the evolutionary benefit of life and therefore the shape and length of the curves below. There is also wide agreement that reproductive decline with age is a symptom of aging. A non-aging organism would also not display decline in its reproductive ability.

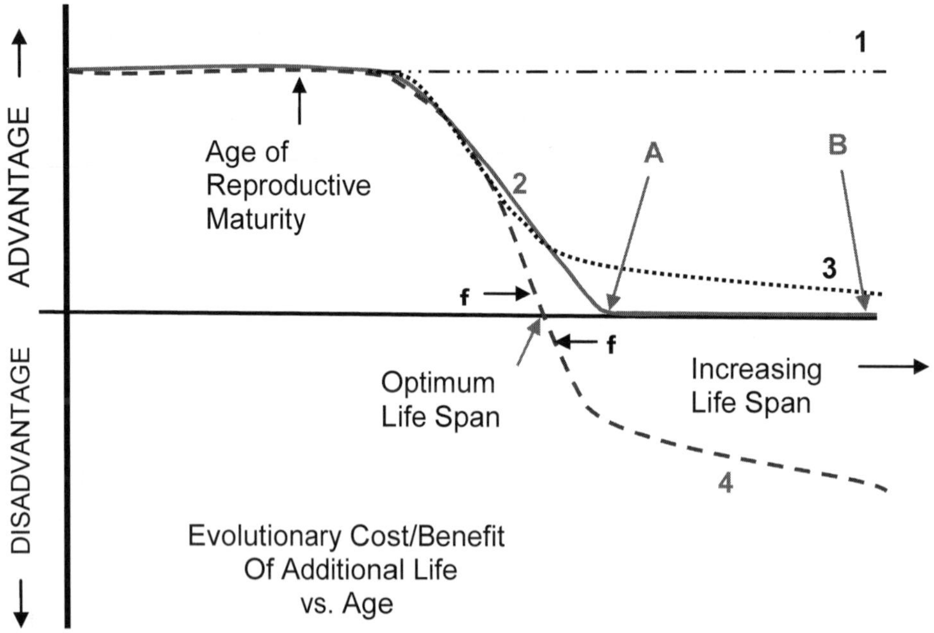

Figure 14 Evolutionary Value of Additional Life vs. Age

The scientific disagreements concern the later (older) portions of the curves during which aging occurs.

Darwin, (interrupted horizontal line, concept 1), did not suggest that the evolutionary value of survival and reproduction varied with organism age. Any incremental increase in lifespan added to an organism's opportunity for reproduction and therefore created evolutionary benefit that continued indefinitely. The force of evolution was therefore toward development of immortality including the ability to reproduce indefinitely. It was immediately noticed (~1860) by Darwin's critics that most organisms were not in fact immortal and that lifespans varied greatly between otherwise very similar organisms. Further, some species died immediately after reproducing for the first (and only) time. These observed conflicts with Darwin's idea eventually led to development of the other three concepts.

The idea that evolutionary force does not vary with age leads to the idea that aging is the result of fundamental limitations. If the force of evolution has, for billions of years, been toward achieving ever longer reproductive lifespans, it is a reasonable assumption that aging is a problem that the evolution process has been unable to solve. Generic damage and fundamental limitation theories such as the wear and tear theory of aging are based on concept 1.

Peter Medawar (solid line, concept 2) proposed in 1952 that the evolutionary benefit of additional life in mammals becomes so negligible as to have no evolutionary effect at some species-specific age linked to reproductive maturity. Genetic drift could then introduce random changes that cause aging as long as their negative effects only occurred subsequent to that age. A yet longer lifespan has zero evolutionary value but also no disadvantage. His argument was that few wild animals live long enough for aging to become a problem and the few that do have relatively little effect on the evolution of a population. According to Medawar, a wild population of immortal animals would be very similar to a population of aging animals, would have a similar age distribution, and would evolve in *exactly* the same way. The mutation accumulation theory of aging is based on concept 2. Medawar's idea relates *intrinsic* limitations on lifespan (aging) to *extrinsic* limitations (predators, food supply, environment, etc.). He assumed that all species are subject to natural selection and that therefore for each species there is an age beyond which extrinsic limitations would so dominate that intrinsic limitations would have no effect. Further, Medawar's theory is population-oriented as opposed to individual-oriented. He explicitly assumed that an older immortal organism would be functionally identical to its younger self and therefore have identical fitness. The difference between older and younger immortal organisms is that there are fewer of them because of extrinsic causes of mortality.

Many other proponents of non-programmed aging (e.g. G. Williams, T. Kirkwood) subsequently proposed (dotted line, concept 3) that the evolutionary value of additional lifespan free of the deleterious effects of aging declines but never declines to exactly zero. A longer life would allow progressively more opportunity for reproduction and consequently at least some advantage in the propagation of an individual organism's design. In addition, aging causes degradation at relatively young ages and this degradation has obvious negative effects on survival potential. These theorists therefore propose that aging must be an unavoidable adverse side-effect that is coincidentally *rigidly linked* to some beneficial design property. Because the evolutionary benefit of life declines once an organism has had some opportunity to reproduce, the ultimately catastrophic disadvantage of aging could be outweighed by a relatively smaller compensating advantage to younger animals. The assumed rigid linkage prevents the evolution process from producing a design that accomplishes the benefit without the adverse side-effect. The rigid linkage concept is itself subject to counter-argument as described in Chapter 7.

Note that this idea is a subset of Medawar's idea. The ***net*** evolutionary effect of aging is *effectively zero* if we sum the minor negative effect of not living longer (curve 3) and the positive effects of the assumed linked properties. If we look at lifespans of thousands of different species it is obvious that a species can evolve whatever lifespan it needs from an evolutionary standpoint. Therefore if the net value of living longer and reproducing more at some age were even slightly positive it is assumed that the organism would have evolved a longer life. The reason for the term "effectively" is that it is assumed that the evolutionary advantage conveyed by some evolved trait must exceed some tiny threshold for the trait to be evolved and retained. We all have eyebrows therefore we infer that eyebrows must convey some tiny evolutionary advantage.

All of the above concepts are compatible with traditional (c ~1950) evolutionary mechanics theory, which requires evolved traits to increase the ability of *individual* organisms to *survive* or reproduce. Concepts 2 and 3 provide a much better fit to multi-species lifespan observations than concept 1.

Finally, advocates of purposely programmed or adaptive aging such as Weismann and subsequent current proponents (dashed line, concept 4) contend that beyond some species-specific lifespan, also dependent on age of reproductive maturity, additional lifespan creates an evolutionary *disadvantage* and that therefore organisms evolved mechanisms for proactively limiting their lifespans to achieve an *optimum* lifespan. In this case there would be evolutionary force (f) to both achieve the species-specific optimum lifespan by means of myriad complex evolved survival characteristics and also to avoid exceeding it by means of an evolved lifespan limiting mechanism. Because, unlike the other concepts, there is evolutionary force toward *limiting* lifespan, there is an evolutionary rationale for the development of a complex mechanism to accomplish the limiting function.

This idea is actually a relatively minor extension of Williams' idea. Instead of proposing that at some age the net value of further survival and reproduction declines to zero it proposes that at some age the net value of surviving and reproducing becomes at least slightly negative.

Programmed theories of aging (complex lifespan regulation) based on life-value concept 4 also fit the multi-species lifespan observations (see following table) and, in addition, provide a better fit to many other observations including the following:

-Caloric restriction effect toward increasing lifespan

-Stress effects toward increasing lifespan

-Progeria and Werner syndrome

-Aging genes produce aging with no known traditional (individual) benefit

-Negligible senescence, apparently non-aging organisms

-Observed lifespan regulation in simple organisms (Kenyon, et al)

-Octopus suicide mechanism (Wodinsky, 1977)

-Similarity in aging symptoms between short and long-lived species

Because the assumed beneficial effects of limiting lifespan could trade off against the presumed residual beneficial effect of living longer shown in curve 3, programmed aging theories do not have to assume the existence of rigid linkages that improbably only affect aging.

All of the evolutionary non-programmed theories (curve 2) assume that there is no evolutionary disadvantage to living longer than the lifespan needed by the organism (point "A" in the diagram). Point "B" or any lifespan to the right of point "B" including effective immortality would also satisfy the requirements of their theory. How then to explain why each organism ended up with just the particular lifespan (point "A") needed by that species and the consequent huge variation in lifespans? If some ancestor species evolved a long lifespan, would this mean that all of its descendent species would have a lifespan at least as long? In order solve this problem non-programmed proponents have to assume that unopposed "random" processes such as random mutations would degrade the lifespan of a species until it reached point "A" for that species. Major difficulties with this assumption are described in Chapter 10. Since programmed theories assume evolutionary force toward both achieving the target lifespan and not exceeding it, they do not have this problem.

Some theorists contend that for animals that nurture their young, the value of life varies between conception and reproductive maturity. For example, the death of a preadolescent would have more impact on its parent's ability to produce descendants than the death of an infant because the parents would have invested more resources and energy in producing the preadolescent. I have not attempted to portray this in the diagram because it is not germane to the aging and lifespan issue.

There are a number of important points to note regarding the value-of-life controversy:

1. The four concepts span all of the possibilities regarding evolutionary value of life vs. age relative to age of reproductive maturity. Despite 150 years of effort we have been unable to definitively narrow the possibilities. Theories of biological aging are essentially dictated by the value of life issue and each concept has its own dependent set of biological aging theories.

2. This problem is analytically very difficult. Concepts 2, 3, and 4, involve theorists "comparing different values of zero" as follows:

   Concept 2 and 3 proponents say that the net value of extended life is zero but contend that it is impossible that it could be even minutely negative as suggested by proponents of concept 4. Example: a proponent of Medawar's or Williams' or Kirkwood's hypothesis or other non-programmed theory (net evolutionary value of extended life is zero) essentially has to argue that the group, evolvability, or gene-oriented benefit of a purposely limited lifespan has an absolute value less than "effectively zero." Otherwise, the benefit would outweigh the (zero) cost and allow evolution of suicide mechanisms. These arguments tend to resemble arguments as to how many angels can fit on the head of a pin!

3. Members of the science-oriented general public are generally unaware of the value of life controversy and have been taught Darwin's concept (1) in introductory biology material. They therefore logically tend to believe in fundamental limitation theories, which in turn suggest that aging is an unalterable property of life and that consequently our ability to prevent or treat massively age-related diseases may be severely limited.

4. Although details (age of reproductive maturity, etc.) vary between species, the life-value concept appears to be very generally applicable. If one believes in, say, concept 2 for mammals, there is no obvious reason for believing in a different life-value concept in connection with birds or perhaps even plants. However, those proposing dependent mammal aging theories often want to consider only mammals while excluding and ignoring any contrary evidence derived from non-mammal species. If specifically asked, they declare that non-mammals are "irrelevant" to mammal aging but that other mammals are "relevant" to human aging. See Science Note below.

5. ***All*** of the evolutionary programmed and non-programmed aging concepts (curves 2, 3, and 4) agree that at some species-specific age the *net* evolutionary value of surviving and reproducing must be effectively zero.

Summary table: Principal theories of biological aging by controlling evolutionary mechanics theory (traditional individual-benefit-only or alternative, which includes non-individual benefit), controlling value-of-life concept, and degree to which they fit observations:

Goldsmith

Evol. Mech.	Evolutionary Value of Life Concept	Dependent Aging Theories	Empirical Fit
Darwin	1) Value of life does not vary with age	- Generic damage theories, wear and tear theories, fundamental limitation theories	Poor
Trad*.	2) Value of life declines to zero	- Mutation accumulation theory, Medawar, 1952	Better
Trad*.	3) Value of life declines, to non-zero positive value	- Antagonistic pleiotropy theory, Williams, 1957 - Disposable soma theory, Kirkwood, et al, 1975 - Other theories in which aging is an unavoidable adverse side-effect rigidly linked to some individually beneficial property	Better
Alt.	4) Value of life becomes negative beyond species-specific optimum age	- Purposely programmed adaptive aging theories, life span regulation	Best

Science Note – Data Point Exclusion

One of the most sensitive aspects of any scientific analysis concerns exclusion of empirical data from a set. To illustrate, imagine an experiment to determine if some agent extends lives of rats. We compare the lifespans of a group of treated rats to a control group of untreated animals. In the treated group some rats live longer than the median control animal, some don't live as long. The question is whether, statistically, the treated group lives longer. A false positive result could obviously be obtained merely by excluding some of the negative data points. In many cases, such exclusion might be legitimate. Rat "F" died early of a congenital defect. Rat "K" was accidentally killed. Rat "W" died in a fight with another rat. Good science demands extremely careful consideration of the validity of each exclusion. Published results should describe exclusions and their rationales.

This principal is routinely violated in discussions about aging theories. People discussing human aging theories based on wear and tear or other generic or even universal damage mechanism (e.g. random changes) frequently pretend that other species are "irrelevant" to human aging and thus rationalize the exclusion of all of the non-human contrary evidence such as the massive inter-species lifespan differences. Proponents of some "mammal" aging theories based on traditional individual-benefit-only evolutionary mechanics decry the exclusion of mammals from human aging theory but tend to pretend that non-mammals are "irrelevant" to mammal aging and therefore ignore rather massive contrary evidence from non-mammal sources.

It might seem perfectly reasonable that an article about a human aging theory, written by someone only interested in human aging, for an audience that was primarily or exclusively interested in humans (e.g. medical people), would not consider non-human species.

Are any of these exclusions legitimate? One can certainly understand that humans are very different from, say, the tiny roundworm *C. elegans*. *Maybe* this difference justifies exclusion of roundworm data from mammal aging theories? The problem with this is that all of the non-programmed aging theories are based on concepts that are extremely widely applicable to living organisms.

Traditional evolutionary mechanics is specifically proposed as applying to *all organisms*. The value-of-life concepts are also apparently applicable to at least all sexually reproducing organisms having specific lifespans. Therefore, neither evolutionary mechanics nor the value-of-life concepts provide a rationale for excluding roundworm or other non-human data from an aging theory. Good science would demand that the theorist identify the specific rationale used in excluding each item of contrary data.

Advice to reader: When reading papers about aging theory, consider carefully what empirical evidence (see chapter 6) is being excluded and the rationale (if any) presented for the exclusion.

5. Digital Genetics and Evolution Theory

Our knowledge of the mechanics of inheritance has increased enormously since the time of Darwin or even the time of Medawar. This chapter is intended to provide only a brief summary of the aspects of modern genetics that are relevant to discriminating between various theories of evolutionary mechanics and dependent theories of aging. Since evolution involves the modification and propagation of heritable information that directs the design characteristics of organisms, an understanding of the mechanics of inheritance is critical to understanding evolution. Evolution is built upon inheritance.

The genetic mechanisms described here can be found in much more detail in a *recent* genetics textbook such as Lewin's excellent GenesVIII[8] (2004) or even newer Genes X (2009).

Early scientists thought that sexual reproduction involved transmission of a miniature microscopic animal. The animal merely subsequently grew larger. However, what was the source of the miniature animal? Another early theory had it that the miniature animals were nested such that the outermost animal grew larger and then transmitted the remaining nested microscopic animals during reproduction. This scheme would apparently be limited in the total number of consecutive reproductions and did not explain why animals shared characteristics of both parents.

By Darwin's time, it was apparent that what was transmitted in reproduction was primarily ***information*** that enabled the descendent organism to construct itself according to a plan that was provided jointly by its parents. The information was somehow stored in the organism during its life and then transmitted to descendants during reproduction.

An earlier evolution theory known as Lamarckism after originator French naturalist Jean-Baptiste Pierre Antoine de Monet Chevalier de Lamarck (1744 – 1829) held that traits acquired during the life of an organism by use or disuse of a body part could be inherited. If a giraffe stretched its neck reaching for food its descendants would have longer necks. If a blacksmith developed enlarged arm muscles as a result of his profession, his sons would be more likely to have larger arms. This idea, that events that happened during the life of an organism could affect and modify the stored information in a structured way has been subsequently disproved.

Therefore, sexual reproduction involves the *copying* and *transmission* of genetic information as well as the *structured merging* of information from two parents to direct the design of the descendent organism. Growth of an organism involves *reading* and *interpreting* the information and constructing an organism whose design is specified by the plan conveyed by the transmitted information. Finally, the design of all organisms provides some mechanism for *storing* the genetic information so that it is available for subsequent reproduction.

Evolution theory tells us that species evolved from other species so it is obvious that some mechanism must exist for *modifying* genetic information. Evolution theory says that species could build upon and extend the characteristics of ancestor species so that it is clear that the modifications are progressive and cumulative. We now know that evolution of life on Earth has been progressing for about four billion years. The mechanisms that are being used by nature to copy and store genetic information are apparently capable of such high fidelity that such a progression is possible. If there is some limit to the ultimate extent to which evolution can progress, we have apparently not yet reached it.

Analog and Digital Data

Here we need to take a detour to discuss the two ways in which information can be stored and transmitted, namely *analog form* and *digital form*. These two modes for transmission, storage, and copying of information have very different properties. An understanding of these properties is critical to understanding genetics issues in evolution and aging theories.

In Edison's phonograph (1877), a diaphragm converted the pressure of sound in the air into the displacement of a needle that then made tracks on a wax or tinfoil cylinder. The displacement and path of the resulting track was *continuously variable* in response to the sound pressure. The phonograph was an instrument for storing and reproducing information in analog form. The information was both accepted and returned as a *serial sequential stream*. The information stored in such a recording could be copied to make thousands of duplicate recordings that could be transmitted far and wide. AM and FM radio, analog television, audio cassette tapes, and VHS video tapes are examples of current analog data systems.

In contrast, Morse's telegraph (1844) represented a serial *digital* communications system. Instead of being continuously variable, the signal sent down a telegraph wire was *binary* and had only two states, "mark" and "space", known in communications terms as *symbols*. The operator converted written characters into a *code* consisting of long or short marks separated by spaces. Longer spaces denoted the beginnings and ends of characters. Yet longer spaces denoted the beginnings and ends of words. Currently, the Internet, CDs, DVDs, space communications systems, and digital television are all examples of digital communications systems.

One of the problems with analog communications is *noise*. Since the *signal* (the desired, information) is continuously variable, any disturbance introduces an error or discrepancy from the original signal that cannot be removed because it is indistinguishable from the signal. This is an especially severe problem when consecutive copies of information are made. Edison could make thousands of copies of an original because each copy was a copy of the original, that is, there was only one *generation*. If we needed to make a copy of a copy of a copy of a copy the cumulative noise buildup would be very severe. Each generation adds more noise.

In digital systems, noise is not as much of a problem. Because the telegraph had only two symbols, disturbing noise could not cause an error unless it was so great as to cause "mark" to be confused with "space." For the same reason digital data can be *regenerated* and noise removed. Copies of copies are not as much of a problem with digital data. A copy of a CD or DVD is usually exactly as good as the original. Copies of copies can be made indefinitely.

Some digital systems have more than two symbols. For example, the English language, a serial digital communications system, has 27 primary symbols (A – Z and space).

Any digital system can ultimately be reduced to *binary digits* or *bits.* That is, we could convert the 27 possible English symbols to 27 possible combinations of five binary digits. This scheme was used in the *Baudot code* (invented in 1870 by Emile Baudot) used for early automated telegraph (teletype) machines.

Here is an illustrative example of a digital communications system. The relevance of this "engineering" discussion to genetics will soon be apparent.

Suppose we had some automated weather stations and wanted to send a digital message from each station to our central location several times per hour. Suppose further that our system works by sequentially sending any of four possible symbols denoted A, B, C, and D. We could devise a message *format* or *code* as follows:

Format	sss	www	vvv	dd	+tttt	hhh	ppp
Content	Sync	Stn ID	Wind vel	Wind dir	Temp	Hum	Pres
Sample	AAA	ABB	DCA	AA	DABC	CDA	AAB

Symbols would be sent in order reading from left to right.

First, the station sends a three symbol *synchronization pattern* sss. This is a known fixed pattern that allows the receiver to determine the meanings of subsequent symbols. We could choose the value AAA for the synchronization pattern. (In English, synchronization is performed by spaces and punctuation characters.) We can follow this with three symbols (www) denoting the weather station sending the message.

Next are three symbols giving the wind velocity. Since there are four possible symbols (A,B,C,D), three symbols together have a total of 64 possible values. We could convert the analog wind velocity to a number between 0 and 63 and then represent it with three symbols. AAA would correspond to 0, AAB correspond to 1, and DDD correspond to 63. Next come two symbols denoting wind direction. Two symbols have 16 possible values. AA could correspond to North, AB to North-Northeast, and so forth. The total number of possible values and therefore the magnitude of a single "step" or "count" is called the *granularity*.

Note that information is being lost in the conversion between continuously variable analog form and digital form. Although the actual wind direction might be anywhere between say North and North-Northeast, the analog to digital converter is forced to pick one of the allowed values. Presumably, if the actual direction is closer to North than North-Northeast or North-Northwest it picks North. This discrepancy between the actual analog value and the digitized value is called *quantizing error* even though it is not actually an error but rather a fundamental property of digital communications.

Next, we have a single symbol denoting the sign of the temperature, (D denotes positive) followed by four symbols denoting temperature. Since four symbols are used, the temperature can have 256 possible values allowing temperature to be conveyed more precisely.

We then add three symbols each for humidity and air pressure. The system does not care if there are extra junk symbols preceding or following the message as long as they do not duplicate a synchronization pattern. This is because the format or rules for transmitting and receiving the data call for looking for the synchronization pattern and then interpreting only the specified following symbols based on their distance (number of letters) from the synchronization pattern. Notice that all the messages have the same *organization* and format. The information is represented by the specific digital *content*, which varies from message to message.

One difficulty is apparent. If the temperature or some other parameter had the value 0 (corresponding to symbols AAA) then the receiver might synchronize at the wrong place in the symbol sequence causing all the data to be misinterpreted. We could eliminate this problem by forbidding the value 0 in any of the data sequences and digitizing all the temperatures and other parameters to values starting at 1 instead of 0.

This simple system illustrates some of the properties of digital communications systems.

Because there are only a finite number of symbols, all the data in a digital system is ultimately limited in "precision." Nothing is continuously or indefinitely variable. The degree of variability allowed (*granularity*) is determined by the number of different symbols possible (in this case four) and the number of symbols chosen to convey a particular parameter.

In our example, we chose to *quantize* analog information into digital form using equal steps. This was an arbitrary choice. When you speak on the telephone, the telephone company converts the analog amplitude of your voice into a series of digital values between 0 and 255. However, these 255 steps are not equal. The steps chosen are smaller at the quieter (lower amplitude) end than at the louder end of the range. This allows quieter sounds to be represented more precisely.

All digital communications require a *language*. We can define language as all the information that the receiver or retriever of a communication must possess, *in advance*, in order to "understand" or apply the information in a digital message. In our example, language would include the manner in which the information fields were represented or encoded, the order in which various parameters were transmitted, nature of the synchronization scheme, forbidden values, and generally, all the information specified for our example system in the previous several paragraphs. Languages are generally *arbitrary*. We could have specified that the "least significant" symbol be transmitted first instead of last. We could have designed our entire communications scheme completely differently. Different human languages are examples of the arbitrary nature of language.

The consequences of an error in a digital code vary enormously depending on where in the format the error occurs. A single symbol error in the synchronization pattern would cause the entire message to be missed. A single symbol error in the "most significant" (leftmost) symbol of the temperature is 64 times larger than a corresponding error in the least significant symbol. An error resulting in insertion of an extra letter or deletion of a letter in a message would result in misinterpretation of all the subsequent data in the message. Insertion or deletion of letters between messages would have no effect unless a new synchronization pattern was created.

In an analog system, errors (noise) tend to cause minor deviations from the true value of a communicated parameter but all communications have errors. The probability of a deviation is inversely proportional to its size. Bigger errors are less frequent.

In a digital system, error-free communication is much more likely, but errors occasionally still happen. The consequences of a digital error tend to be more severe and less structured. In our example message format there are 22 symbols. An error in which one of the symbols was replaced by an incorrect symbol (a *substitution* error) would cause a major change in reported value unless it occurred in the least significant symbol of a parameter. There are only 5 least significant symbols in our code so more than 75 percent of the possible errors would cause major, even catastrophic, effect. An error in which an additional symbol was inserted or an existing symbol was deleted would be catastrophic in nearly all cases because the subsequent symbols would be misinterpreted.

In modern digital communications systems various methods have been developed to detect and even correct errors. One obvious technique is redundancy. We could send or store the same information three times and compare the data on the receiving or retrieving end. If any of the copies did not agree with the other two we would know it contained an error and discard it. Many more sophisticated ways of ensuring error free transmission and storage of data are in current use.

Another major difference between analog and digital data concerns *structured merging*, which we can define as the combining of information from two or more sources to form a meaningful composite. A fundamental property of analog systems is that it is easy to merge data. Edison's phonograph could record a duet or even an entire choir as easily as it could record a solo singer. The sound pressure variations from the different sources merely added in the air. No change to the recording device was required. Analog photographs can be added to make a double exposure. Analog signals in electrical form can be similarly easily merged by simple addition.

Merging of digital data represents an entirely different and much more difficult problem. Consider the following two digital messages:

```
Build a red brick wall 910 cm wide, 220 cm high and 20 cm deep.
Build a tan brick wall 600 cm wide, 320 cm high and 18 cm deep.
```

There is no way to just "add" the messages to make a composite. We could identify the variable (numeric) parts of the messages, convert from digital to analog, perform addition or averaging operations, and then convert back to digital, a very complex procedure that requires *a priori* knowledge of the specific format of the messages. What if "red" and "tan" was a binary choice with no intermediate possibility?

Another possibility for producing a composite would be to simply replace some characters in the first message with corresponding characters in the second message. For example if we used the first half of message 1 and the second half of message 2 the result would be:

```
Build a red brick wall 910 cm wide, 320 cm high and 18 cm deep.
```

This scheme has some important and severe restrictions. The "format" of the two messages must be the same. If one of the messages had, for example, "high" as the first parameter, and "wide" as the second parameter, the result of the merge would not be meaningful. Second, the length of the two messages must be identical. Suppose the messages were:

```
Build a red brick wall 1000 cm wide, 220 cm high and 20 cm deep.
Build a tan brick wall 600 cm wide, 320 cm high and 18 cm deep.
```

Now the meaning of the merged message would be disturbed because the result would be:

```
Build a red brick wall 1000 cm wie, 320 cm high and 18 cm deep.
```

Functionally the result is similar to an insertion or deletion error.

The reason for this detour is that the "genetic communications system" is in fact a serial digital system and bears an eerie resemblance to modern digital data systems. The genetic system has four symbols, synchronization patterns, formats, redundancy, error detection, merging, framing errors, language, and many other properties of digital systems. The genetic system is constrained by the *digital data* considerations described above. This has significant consequences for evolution theory and aging theory as will be explained in detail. We can use the term *digital genetics* to refer to these aspects of genetics that are driven by the digital nature of the genetic system.

At Darwin's time, many thought (despite some rather obvious discrepancies) that inheritance was an *analog*, continuously variable, averaging process. It was thought that characteristics of progeny tended to average out the characteristics of their parents. An analog method of inheritance would neatly fit casual observations. Variation is not only a fundamental property of an analog system but the occurrence frequency of a variation is inversely proportional to its size. This fits the bell shaped curve we would expect if we measured, for example, height variations in 18-year-old males. Darwin's world was an analog world. Darwin had no reason to consider the digital concepts discussed above.

Gregor Mendel (1822 – 1884) was an Augustinian monk who conducted very extensive crossbreeding experiments with peas and other plants. Unlike Darwin, Mendel followed the inheritance process between specific individual organisms, to their descendants and their descendant's descendants. Mendel's paper *Experiments in Plant Hybridization* (1865)[9] was not widely noted until much later and was unknown to Darwin. Mendel determined that some inherited characteristics were *discrete* or *binary*. That is, there was a minimum unit of inheritance such that some characteristics were either inherited by a given individual, or not, with no averaging or intermediate possibility. Inheritance of traits was not continuously variable. Mendel also noticed that some inherited characteristics were *latent*. Progeny could exhibit characteristics that were not displayed by *either* of their parents but were displayed by grandparents or other ancestors.

Figure 15 Gregor Mendel

Mendel's paper provided the first clue that inheritance was not an analog process. Subsequent extensive research into inheritance disclosed the existence of chromosomes and other aspects of the digital inheritance system.

Watson and Crick in 1953 published their famous paper[10] *A Structure for Deoxyribose Nucleic Acid* describing the basic mechanism (the "double helix") whereby genetic information is recorded, copied, and transmitted in all living organisms. They shared the Nobel Prize in Medicine for 1962 with co-discoverer Wilkins.

Serial Digital Genetic Codes

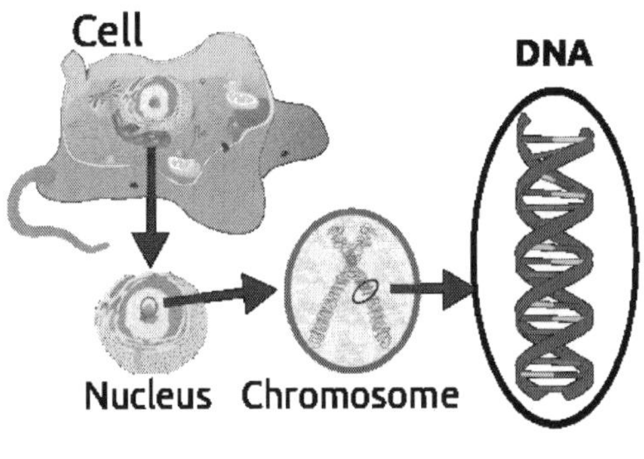

Figure 16 DNA Structure

As determined by Watson and Crick, and extended by many subsequent investigators, the system used by nature to store, copy, and transmit genetic information is a *digital* system. Genetic information is conveyed by the *sequence* in which the organic compounds adenine, guanine, cytosine, and thymine are strung together to make long molecules of DNA. These sequences then ultimately determine all the inherited characteristics of the organism. In communications parlance, this would be a *serial digital code*. Because of the digital nature of the genetic code, some parts of genetic sequences have been faithfully reproduced (i.e. consecutively copied) for *billions of years*. As we have previously seen, an analog system would never be capable of accommodating the very large number of consecutive duplications involved in the evolution of life on Earth.

Since there are four possible *bases*, (A, G, C, and T for adenine, guanine, cytosine, and thymine), each base corresponds to two bits of information. We could translate A to 00, G to 01, T to 10, and C to 11 and then represent any amount of genetic code as a binary number sequence. AGTTC would then be 0001101011. The bases are the symbols of the genetic code.

In binary terms, 0 is the *complement* of 1 and 1 is the complement of 0. The binary sequence 100111001 is the complement of 011000110. The complement of a complement returns the original sequence. Complementing a sequence therefore does not remove any information. In genetic code terms, A and T are complements and C and G are complements. The sequence ATTGCCC is the complement of TAACGGG. The "*double* helix" DNA molecule actually contains a sequence of bases wrapped with the *complementary* sequence, hence the terms “double helix” and “base pair.” The two sequences redundantly carry the same information. From an information viewpoint, a “base”, is the same as a “base pair, is the same as a “letter”, is the same as a “nucleotide.”

The Human Genome Project in 2001 released a preliminary report describing the actual sequence of the genetic content (or *genome*) for humans and determined that the human genome contains about 3.3 billion bases of information. By early 2003 the sequence had been 99.9 percent determined[11]. In computer terms this is about 6.6 billion bits or 825 megabytes of data – small enough to fit on your laptop computer’s hard disk. Approximately half of the genome consists of *repeat sequences* that are highly repetitive and therefore, according to information theory, contain very little information. Some of the repeats are *tandem repeats* that consist of sequential repetitions of a simple sequence (e.g. ATATATATAT...AT). A large amount of the remaining code has no obvious function. Although the sequence has been determined, the actual specific functions of most of the genetic code remain unknown.

Genetic code in more advanced organisms is transmitted in the form of contiguous, sequential, DNA molecules called *chromosomes*. Chromosomes are visible under certain conditions using optical microscopy and were discovered by Walther Flemming in 1882. In 1907, Thomas Hunt Morgan associated chromosomes with inheritance using fruit fly experiments.

Humans have 23 chromosomes. Mice have 20. Dogs have 39. Some plants have more than 100. Small objects in the female egg cell called *mitochondria* that are duplicated in subsequent cells transmit a small percentage of human DNA.

Chromosomes have special sequences on either end called *telomeres* (in humans, repeats of the sequence TTAGGG). Another special sequence more centrally located (position varies depending on the chromosome) is called the *centromere*. When a cell divides to form a second cell, the genetic information content is duplicated in a process called *mitosis*. Chromosomes in a cell are normally in an extremely compact spherical shape. During mitosis, chromosomes expand to a somewhat less compressed form in which they can be seen as the familiar microscopic rod-shaped objects. If completely unwrapped and extended, the chromosomes in a single human cell would total several cm in length. The telomeres, centromeres, and other structural aspects of chromosomes are known to be essential to the proper duplication of one (and only one) complete set of chromosomes during cell division.

Almost every non-sex cell in more advanced organisms has two sets of genetic data, (two sets of chromosomes) one inherited from each parent. In this *diploid* configuration, the chromosomes are paired, that is, corresponding chromosomes are physically attached to each other to form the familiar conjoined rod shapes. Sperm and egg cells only have one set of chromosomes in a *haploid* configuration.

Bacteria do not possess paired chromosomes. Instead, their DNA is typically in the form of a single, much simpler, loop.

Because advanced organisms have two sets of data (about 1.6 gigabytes in humans) the inheritance, (genetic data communications) process in advanced organisms is substantially different from that of simpler organisms.

Mice have a genome of about 3 billion bases (only 10 percent less than humans).

Yeast has a genome of about 12 million bases, 6000 genes on 16 chromosomes.

The bacteria *e coli* have a genome of 5.6 million bases.

A major genetic curiosity, the microscopic amoeba (*Amoeba dubia*) has 670 billion bases in its genome!

To further illustrate the information content, the upper and lower case characters in the English alphabet (52 alphabetic characters, 10 numerical digits, and space) could be represented in binary form using 6 bits per character. The phrase "Four score and seven" would correspond to a binary string of 120 bits and could be expressed in genetic code using 60 bases. (This book contains about 300,000 characters equivalent to 900,000 bases of genetic code or 16 percent of the data in an *e coli* genome.)

The probability of duplicating "Four score and seven" by random combination of bits (as might be done if you had "enough monkeys and typewriters") is one in 2^{120} or one in 10^{36}. It would take a very, very, large number of monkeys and typewriters a very long time to randomly duplicate even this very short phrase! A single error might result in "Four scBre and seven." Several errors could look like "Fouw scory and 8even."

The reason for this diversion is that geneticists can trace descendancy at the species as well as the individual level. If you and any other living thing share a significant sequence of code that is approximately the same then you and the other organism must have had a common ancestor because the chances of a random duplication are impossibly low. Not only can they determine if you are related to your alleged children, they can determine if mice and men had a common ancestor (yes, of course) and can even determine from the number of errors that have crept into the genetic messages approximately how long ago humans and mice had a common ancestor (about 50 million years ago). It is possible for DNA in an organism to be, in effect, "cross-contaminated" with DNA from another organism but this method is considered minor relative to direct inheritance of DNA sequences.

Errors and Mutations

Errors introduced in copying or storing genetic data are the source of the genetic changes that drive evolution. Some errors, such as in a sequence which controls basic cell design, or oxygen transport, or other crucial process, are almost always immediately fatal and so are immediately "selected out" and do not propagate into the genetic code of descendent organisms. This sort of sequence tends to be "well conserved" after billions of years. Humans share some sequences with yeast that both humans and yeast must have received from a common ancestor. Other sequences that control "how much" (how long a claw, how much fur, etc.) are the source of the variation that drives natural selection. An error in such a sequence might only cause slight variation of a parameter and only very mildly affect fitness. Finally, some sequences (possibly more than 90 percent of the human genome) have no apparent biological purpose. Changes in such a sequence generally have no immediate effect on the organism and are putatively not selected against at all, thus apparently freely propagating to future generations. Since larger animals have trillions of cells, there are trillions of opportunities for mutations. However, for a mutation to be inherited it must occur in the sequence of cell division between the original egg cell and the subsequent egg or sperm cell.

In modern electronic data systems, it is not unusual for errors to occur more or less frequently depending on the *pattern* of the data. Errors in both electronic and genetic systems can be caused by *substitution* of an incorrect letter in a sequence and can also be caused by *deletion* of a letter or *insertion* of an extra letter.

In the genetic code, which is all about pattern and sequence, it is not surprising that it is also true that the chance for an error is pattern sensitive. For example, humans have a genetic structure called a *variable number tandem repeat* (VNTR). Copying errors (insertion/deletion errors) which change the length of these repeats are thought to occur virtually every generation. (These are the sequences whose lengths are compared in some types of forensic genetic fingerprinting.) Another illustration of *pattern sensitivity* is the *restriction enzyme*. There are many different enzymes which can cause strands of DNA to be physically broken at points where a particular sequence exists. For example the enzyme *sgf I* causes breaks where the pattern GCGATCGC is encountered. Because of pattern sensitivity, the probability of particular errors varies enormously and is difficult to predict.

In the genetic code, occasionally sequences are duplicated. Genes in the duplicated sections can have subsequent errors that sometimes result in new, useful genes. Presumably, this is the mechanism whereby a more complex and longer genome can evolve from a simpler one.

In human genetic code there is a specific pattern of about 300 bases called the *alu* element. Alu appears about one million times in the human genome and is thought to have a significant role in affecting duplications, which in turn, have a significant role in genetic diseases as well as in implementing evolution of the genetic code. Alu elements represent about ten percent of human genetic code, have no known biological function, and are often considered part of "junk" DNA.

Genes

Genes perform the actual control of physiological functions. Each chromosome can have thousands of genes. The human genome contains approximately 30,000 genes but the actual number is still unknown.

The structure of the sequence of information representing a gene as seen reading sequentially along a chromosome typically includes *regulatory regions* at the beginning or end of the gene sequence that determine when and where the gene is activated.

A gene is often thousands of bases in length. The *coding region* determines which protein will be produced by the gene, that is, the sequence of amino acid molecules which will be constructed to produce a particular protein molecule (often referred to as the *gene product).*

The properties of a protein are determined not only by the number and type of the amino acid molecules used in its construction but also by the particular sequence in which the amino acids are assembled. The long protein molecules tend to "fold up" in very complex ways depending on the particular sequence. This folding and consequent *shape* of the molecule affects its properties. There are therefore an essentially infinite number of possible different proteins.

A particular three-letter sequence, ATG, is the synchronization pattern denoting the start of a coding sequence; other three letter sequences (known in genetics parlance as *codons*) denote particular amino acids to be sequenced into a protein and the end of a coding sequence. Since there are 20 possible amino acids and 64 possible codons, some errors in the third symbol of a codon have no biological effect. For example, CTA, CTG, CTT, and CTC all code for valine. This is a form of redundancy.

The regulatory regions determine when, where, and how much product will be produced. Some products are only produced in the liver; some are produced only at certain times in an animal's life, and so on. The regulation involves the detection of *chemical signals* which can either enhance or inhibit the gene's *expression.* Although some genes produce proteins used in the construction of tissue, many, probably a majority, produce products that act as signals to activate or inhibit other genes thus allowing the construction of a very complex regulatory logic framework.

If the regulatory region determines that a gene is activated, the cell starts making copies of the genetic information in the coding region in the form of small RNA molecules with sequences corresponding to the coding region. These *messenger RNA* molecules are used as templates by the cell machinery that produces the proper protein molecules. (Sometimes the RNA molecule itself is the gene product and performs some biological function such as acting as a signal to other genes.)

The RNAs will preferentially adhere to a complementary string of code. "Gene chips" carrying hundreds of samples of potential RNA complements can be used to test for the presence of specific RNAs in a sample. Using such gene chips, researchers can detect the presence of various different RNAs in various tissues and thereby determine which genes were activated. In connection with anti-aging research, detecting the differences in gene activity between a caloric restricted animal and not, or between a progeria victim, and not (see next chapter) could produce valuable clues regarding aging mechanisms.

We can think of a specific "gene" as a message defining a product that accomplishes a particular biological function. Since all multi-cell organisms have a common basic cell design and function it should be no surprise that there are genes that are common to all such organisms. As organisms become more similar they share more commonality. It is estimated that 99 percent of mouse genes have an equivalent human gene that produces a very similar product.

Genes represent a complex digital data structure. A large proportion of the possible random changes to a gene result in its function being destroyed, that is, inactivation of the gene. This has significance to the process of evolution.

The *organization* of the genes in the genome tends to be very different between even similar species. Mice have a different number of chromosomes from humans and the equivalent genes are generally in a different order on different chromosomes. Some genes are organized in groups or *clusters* that are conserved between mice and humans.

Coding regions in the genes of more complex organisms have *introns*. Introns are portions of the coding regions of complex organisms that are *spliced out* and deleted from the code during the creation and processing of an RNA molecule. The deletion is caused by patterns at the beginning and end of the intron that match in a particular way. Since the introns are deleted, they have no known biological effect and are often considered "junk" DNA. The remaining (functional) portions of a coding region which are expressed in the RNA and subsequent protein are called *exons*. Exons are thought to represent only about one percent of human DNA data while introns represent about five percent of DNA. Human genes have an average of five introns and a maximum of 178 introns. Bacteria have very few introns.

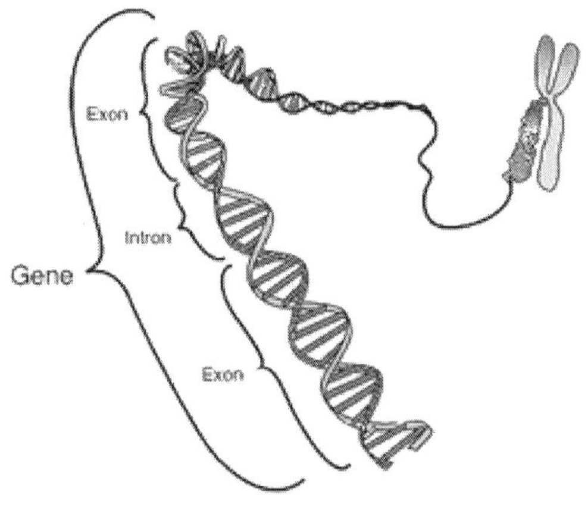

Figure 17 Genes and Chromosomes

Genes are autonomous data units. They contain their own synchronization patterns and operate somewhat independently. Junk DNA can therefore exist between genes without disturbing their operation. The position (or *locus*) of a gene within a chromosome or on a particular chromosome generally does not appear to affect the functional operation of the gene. (In communications parlance such an autonomous data unit would be referred to as a *packet*.) (Some specific genes must be located on the sex chromosomes in order to accomplish sex differences between organisms.)

If we inject a small loose string of DNA containing a single gene into a cell, the cell will happily produce the gene's protein product. This approach is used in some forms of gene therapy. However, the loose strand of DNA would not be duplicated during cell division because such duplication requires the gene to be part of a chromosome. Methods for inserting new genes into chromosomes have been developed and are used in genetic engineering. Such a gene would be propagated during cell division and even possibly during reproduction of the organism. While junk DNA and gene location do not affect the functioning of genes they may well have significant evolutionary effects to be described.

All normal humans are thought to have the same genes, specifying the same or nearly the same products, in the same order, on the same chromosomes. Genetic differences between humans are expressed in the exact digital content of their genes, generally minor differences such as single letter substitutions.

Mendelian genetics considers that some genes in a particular species can have two different specific data contents or *alleles* such that two different results occur. Often one allele is represented by a gene that is disabled and therefore produces no functional product, while the other allele is represented by the functioning gene, a binary situation. In practice, some genes can have more than one functioning state and a single gene can therefore have more than two alleles. A complex gene having tens of thousands of bases could possibly have many alleles.

A single substitution difference in a coding region exon (for example an A could be replaced with a T) could cause a different protein or RNA product to be produced, which in turn could have a significant effect but could also have a mild or negligible effect. An error in the regulatory region or an error that deletes the start codon or adds a stop codon could cause the gene to become disabled and produce no useful product. An insertion or deletion in a coding region is likely to disable the gene because all subsequent data would be misinterpreted. The insertion or deletion of exactly three contiguous letters might well have only a minor effect because it would only cause an extra amino acid molecule to appear in the resulting protein (or a single amino acid to be deleted). Other errors could have more minor effects such as changing the amount of product produced. Many of the more than 1000 known human genetic diseases as well as most of the normal variations between individuals are caused by such single letter differences in the genome.

In many cases of genetic disease, if one parent's gene is disabled, the other parent's corresponding gene provides enough product so that significant symptoms are avoided. The child and the first parent are *carriers*. If the genes received from both parents are defective, then the child has the *recessive* genetic disease. If one gene does not provide enough product to avoid symptoms, or if an incorrect and deleterious product is produced, then a defect in either parent's gene can cause disease symptoms in a *dominant* genetic disease or other trait.

Many human genes appear to be duplicated, another form of redundancy.

Therefore, by far the most likely possibility in a mutation is a single letter error. It would appear to be ridiculously unlikely that an entire new functioning gene could be produced by a random mutation. The significance of this is covered in the section on aging genes.

Since the sequence of the human genome has been completed, it might seem a simple matter to have a computer program search through the genome, and identify genes by their characteristic data patterns such as start and stop codons, regulatory sequences, and intron patterns. In practice, although the start and stop codes are definite, the patterns involved in regulatory sequences and the patterns that denote the borders of an intron are often quite vague in that many different patterns appear to accomplish the same result. In addition, the genome contains *pseudogene* patterns that resemble genes but are not functional. A pattern can be "definitively" considered a gene if a gene with the same or similar exons has been found in another species, or if a genetic disease or other trait has been traced to two different forms of the (otherwise) same pattern. Because of these difficulties, we do not yet know for certain even how many genes are in the human genome and have "definitively" identified relatively few genes.

About half of the human genome consists of repeats of very short (2 – 5 bases) or relatively short (<300 bases) sequences. Since these repeats and other "junk" DNA are between genes or in introns they have no apparent effect on an organism's functional design. However, they do have an apparent evolutionary effect in that they influence mechanisms that cause segments of code to be duplicated, copied to another part of the genome, or deleted.

Introns appear to have a similar evolutionary effect. The sections of expressed genetic code (exons) between introns appear in some cases to correspond to "building blocks" or "modules" that have been used by nature to produce a family of different proteins each of which consists of one or more common modules added to a unique sequence. Although the content and length of introns in a particular gene tends to vary between species, the exons and the number of introns tend to be more nearly conserved.

Meiosis and Recombination

As mentioned, (haploid) sex cells have only one copy of the chromosomes so that when a sperm and egg cell are united the resulting (diploid) cell and subsequent cells have a normal complement of two sets of chromosomes. In order to do this, half of the genetic material is not used during the creation of a sperm or egg cell. This process, called *meiosis*, and other aspects of sexual reproduction are extremely complicated as will be summarized below. The purpose of this section is to demonstrate the enormous difficulty nature has endured in order to produce the maximum possible genetically transmitted and structured *variation* in organisms despite the digital nature of the genetic code. These extremely complex evolved mechanisms further validate Darwin's theory of natural selection by means of natural variation and also lend credibility to certain adjustments to Darwin's theory as well as adaptive theories of aging as will be explained in Chapter 7.

In the process of meiosis, one chromosome from each set of two coming from the two parents is randomly selected for transmission in the sperm or egg cell. Humans have 23 different chromosomes (designated as numbers 1 through 22 in order of decreasing length (as seen in the rod form) and either "X" or "Y"). Note that the complex *recombination* mechanism has to guarantee that exactly one of each set of two chromosomes will be transmitted and that we do not possess three of chromosome 1 and none of chromosome 2, etc.

We can illustrate the effect of recombination as follows:

Suppose that a single pair of parents could produce 1,000 children. All of the children, (excepting identical twins), would be different from each other, different from their parents, and different from their ancestors. Each child contains two sets of genetic data derived by randomly merging the four sets of genetic data possessed by the parents. Each of the four sets of data possessed by the parents is only very slightly different from the other three. However, because of the "cascading" effect of combining the data variations in different ways, the range of differences between the 1,000 children will be greater than the differences between the parents. For example, we would expect to find some children that are shorter than either of their parents and we would also expect to find some children that are taller than either of their parents. Recombination, unknown to Darwin, fundamentally alters the process of evolution as will be described.

The differences produced by recombination are constrained by the differences in the original four sets of genetic data. If, for example, both parents were tall, blue-eyed, blond, Scandinavians, descended from generations of blond, blue-eyed Scandinavians, we would expect all the children to be relatively tall and blue-eyed. If one of the parents was such a

Scandinavian and the other was a short, darker skinned person from a distant geographic origin, we would expect the variations between children to be much greater.

Suppose it was somehow possible to mate identical twins of some species. Would all their descendants be identical? Although the twins have identical genetic data, they each have two different sets of data. Their descendants are the result of recombining these two sets in different ways and will therefore be different from each other. If each twin possessed two identical sets of genetic data, *then* their descendants would be identical.

Crossover

A random sort of human chromosomes would result in 2^{23} or 8,388,608 different possible combinations. *Each* parent performs such a random shuffle of the chromosomes received from their parents in producing the sperm and egg cells. This would appear to guarantee plenty of variation. However, it was eventually determined through inheritance studies that reality was actually yet more complicated. If only the chromosomes were shuffled, then inheritance of a gene on a chromosome would be tied to inheritance of another gene on the same chromosome. If you inherited one gene from one grandparent, you would have to also inherit the other gene from that same grandparent. (This would make it impossible for nature to sort out the beneficial or adverse effects of different mutations on the same chromosome and therefore drastically limit the process of evolution.) At the same time if the two genes were on different chromosomes, inheriting one would be completely independent of and not affect the chance of inheriting the other because of the random chromosome shuffle.

(The plant traits that Mendel used in his experiments happened to be on different chromosomes. (Plants tend to have many chromosomes.) If this had not been the case he would probably still be trying to make sense of the inheritance patterns as explained below!)

Geneticists discovered that if traits were controlled by genes on different chromosomes the inheritance pattern was, as predicted, completely independent. However, if genes were on the same chromosomes the inheritance of the respective traits ranged from almost independent (inheritance of one trait was random relative to the other) to nearly totally dependent (inheritance of one trait almost always meant inheritance of the other). They deduced that during construction of sex cells (meiosis) one or more contiguous segments of a parent's chromosome is exchanged (*crossed over*) with the other parent's chromosome to make a new chromosome that is a data composite of the two parents. The length and position of the swapped segment is almost random. As a result, *the probability of inheriting any two genes on a single chromosome from one parent is proportional to the physical "data" distance (number of bases) between the genes on the chromosome.* If two genes are physically close, then they almost always would be inherited together, if physically distant, their inheritance would be almost independent. Using this *genetic linkage* (also sometimes referred to as *genetic distance*) principal and mind numbingly tedious inheritance studies, geneticists have been able to determine the approximate physical chromosome location ("locus") of many genetic disease genes. For example, it was determined that Duchenne muscular dystrophy is caused by defects that disable a gene located near the middle of the short arm of the X chromosome. This gene produces a protein, dystrophin, which is needed for proper muscle function. (The genetic distance approach is very difficult if the disease or trait is the result of two or more genetic differences.)

Unequal Crossover

Initially, it was thought that the crossover mechanism exchanged segments of identical length located at identical positions in their chromosomes. A typical chromosome might contain 100,000,000 bases of data. A crossover could involve exchanging exactly 31,500,354 bases starting at base 15,213,655 on each chromosome as measured from the beginning of the chromosomes.

If true, this arrangement would represent a major limitation on the process of evolution. Suppose a mutation to an individual organism caused an *insertion* or *deletion* of a single base at position 136. Now remember our earlier discussion regarding merging of digital data and the sorts of errors that are created when attempting to merge data strings of different lengths. An insertion at letter 136 would cause one of the crossed over segments in the above example to start one letter further along the chromosome than the swapped segment, causing what amounts to a deletion of one letter. At the same time an insertion of an extra letter would occur at the end of the swapped segment. Any subsequent mating attempt resulting in crossover between a chromosome that had the insertion/deletion and one that did not would result in at least two additional errors because the beginnings and ends of the swapped segments would not match. *Each* subsequent mating and crossover would cause additional errors to occur. Catastrophic disruption of genetic data would be rapid. Therefore, mutations that caused insertions or deletions would be essentially infeasible under the equal crossover arrangement. Insertions appear to be essential to the creation of more complex genetic data and therefore to the creation of more complex organisms.

Eventually it was found that the swapping mechanism apparently only exchanges sequences of data that are nearly identical at least near the ends of the cut sections. A swapped segment of data can therefore contain an insertion or deletion and still not result in disrupting the overall data scheme while undergoing digital merging. To simplify, a gene could be swapped with a corresponding gene even though one gene had a longer or shorter data length than the other.

This *unequal crossover* mechanism apparently depends more on data pattern similarity than position measured from the start of a chromosome so insertions or deletions prior to the swapped section also do not cause data disruption. Unequal crossover helps ensure that genes are only exchanged with corresponding genes and the descendent does not end up with two of some genes and none of some other genes or inherit partial genes or genes with insertion or deletion errors.

In addition to being able to *accommodate* insertions and deletions, the unequal crossover mechanism acts to *cause* insertions and deletions in genetic data. A crossover error can occur if the end of a cut occurs at a place where two identical data sequences (such as two alu elements or a long tandem repeat) occur in a relatively short stretch of data. The cut might exclude (delete) or duplicate (insert) the section of code between the identical sections. It is thought that some genetic diseases as well as variable number tandem repeats are in fact caused by these kinds of errors in the crossover process.

The unequal crossover mechanism appears to be critically important to the evolution process in more advanced organisms by allowing addition to the data content of genes and duplication of genes. Other pattern sensitive mechanisms called *transposons*, also act to transfer genetic data.

Occasionally, humans inherit three copies of chromosome 21 instead of the normal two copies. As a result, 50 percent more of some gene products is made than normal. This in turn results in a genetic disease characterized by mental retardation and physiological abnormalities known as Down syndrome. Other chromosome abnormalities include inheriting less or more than two of any chromosome, swapping of genetic material between different chromosomes, or losing parts of chromosomes. Most such abnormalities cause fetal death or degradation so severe that propagation in a wild population would be impossible. However, the fact that the Down syndrome is not immediately fatal despite duplication of hundreds of genes illustrates that duplication of some genes might happen without severe adverse consequences. (Chromosome 21 is the shortest non-sex chromosome and so its duplication has less impact than other duplications.)

Duplication of genes is part of the process whereby organisms evolve more complexity. In addition, exchange or copying of data between chromosomes must occur because highly related species have the "same" genes on different chromosomes.

Sexual Reproduction

Sexually reproducing organisms have special chromosomes to help manage sexual reproduction. Female humans have two X chromosomes (which are paired and swapped during meiosis like other chromosomes). Males have an X and a Y chromosome. Therefore, progeny always inherit an X chromosome from their female parent and have a 50 percent chance of inheriting an X chromosome from their male parent thus resulting in a 50 percent chance of being either male or female. In humans, the X chromosome is larger and has more genes than the Y chromosome. The gene that triggers "maleness" is on the Y chromosome.

One aspect of this arrangement puzzling to geneticists was how do females avoid having something like Down syndrome? Since females possess two copies of chromosome X and males only have one copy, females would appear to have 100 percent more of some gene products than males (or males have 50 percent less than females). (I know that at this point, some men and woman readers will be saying, "that explains a lot" about women or men respectively.)

Eventually, it was determined that, in females, one (and only one) of their two X chromosomes is randomly "inactivated" such that, functionally, females only have one X chromosome. X inactivation is another in a long list of evolved complexities associated with sexual reproduction.

At least in higher animals, a process similar to X inactivation inactivates certain genes as development proceeds. As stem cells differentiate into more specialized cells, some genes are marked as inactivated (*genetic imprinting*) which partly enables the capability for structural and functional differences in different body cells. The inactivation state of genes is copied during mitosis. So although almost all your cells have all your genes, in most cells some genes are inactivated. This inactivation is removed during meiosis and also when cloning animals from differentiated cells such as skin cells. From an information standpoint, this sort of imprinting has some consequences. Each human diploid cell apparently not only contains the 1.6 gigabytes of genetic data (the same in each cell) but also contains as much as 30,000 bits of "inactivation status data" (differing from cell to cell).

Immunity

Non-sex cells in animals nominally contain the same genetic data but there is an exception. When vertebrate white blood cells are made from precursor cells in bones or thymus gland, some of their DNA is rearranged in a semi-random manner. White cells in an individual organism therefore have slightly different genetic data and slightly different design and function from each other. These differences cause them to respond differently to a specific infecting pathogen protein. Some white cells will detect and attack a particular pathogen, others will be sensitive to other pathogens. White cells that detect a pathogen also are stimulated to divide by simple cell division such that the new cells have the same genetic design and are tailored to the same pathogens as the parent cell. Exposure to a pathogen therefore causes the population of white cells capable of attacking that pathogen to increase. If the animal survives the initial infection, the larger population confers some degree of immunity.

Polymorphism

A *polymorphism* is a situation in which a characteristic possessed by individuals in the normal population varies. If 90 percent of the flies have black eyes and 10 percent have red eyes, this would be an eye-color polymorphism. "Normal" is often defined as meaning that at least one percent of the population has the variation. Most genetic diseases are not "normal."

At the genetic level, it is now estimated that normal humans possess genetic codes that are as much as 99.9 percent identical. However, there are an estimated 3 million places in the human genome where a single letter is different in some cases. In such a location, a letter might be "A" in 85 percent of the cases and "G" in the remaining 15 percent. These "Single Nucleotide Polymorphisms" or SNPs represent and convey the "normal variation" between humans. Presumably, the number of polymorphisms would be much less if we considered only individuals in a particular race and would be progressively still less if we considered only a particular ethnic group, clan, tribe, or family. Extensive research is now under way to identify which SNPs are associated with which identifiable characteristics and to determine how SNPs vary between races, ethnic groups, and families.

If indeed there are 3 million SNPs, (some estimates are as high as 10 million), then there are $2^{3,000,000}$ possible combinations of those SNPs, a very, very large number of combinations. Every human is therefore unique and contains a different combination of SNPs in each of his or her two sets of genetic data. At the same time, as explained earlier, the probability of inheriting any part of the genetic code with some other part depends on the physical data distance (genetic distance) between the two segments on a chromosome. SNPs that were close together on a single chromosome would tend to be inherited together. SNPs that were very close together would tend to be inherited as a unit that would tend not to be divided even after many generations. SNPs that were far apart or were on different chromosomes would be shuffled in nearly every individual. Many SNPs are known to be clustered and their inheritance is therefore complicated. These details regarding the inheritance of variation are critical to theories of evolution such as the selfish gene theory.

Digital Variation

Genetics reveals that the "natural variation" between members of a species actually has *two* sources in more advanced (diploid) organisms:

1. Random changes in genetic data occur in randomly selected organisms that are geographically randomly distributed. If a change causes a minor biological effect, it might propagate widely geographically from its point of origin and eventually become present in a substantial part (but not all) of the species population. Changes cannot significantly propagate to other, co-existing, species. (A change might eventually propagate to encompass essentially the entire population but that change would no longer represent "variation.")

2. Merging or recombination of digital genetic data occurs and creates individuals expressing new *combinations* of *existing* data. The merging process is very complex and has many aspects that plausibly affect the process of evolution. The magnitude of a variation caused by recombination is larger than that caused by individual underlying mutations as explained below. This has importance for evolution theory as described in Chapter 7.

The magnitude of variation produced by recombination is larger than that produced by mutation as follows. The SNPs in genetic data are each presumably the result of a single independent random mutation that occurred in a random individual in a random geological location and are each a binary choice. A set of genetic data found in an individual after the mutation occurred either contains or does not contain the mutant SNP. If an expressed trait variation (such as red eyes vs. white eyes) is controlled by only one SNP, then the range of variation in the population is equal to the variation caused by the single mutation (red or white). Traits that have only a few discrete values such as blood type or Rh factor similarly must be controlled by only a few SNPs.

A trait that appears to be continuously variable (such as height or most "survival" characteristics) must be controlled by many SNPs. Suppose 18 SNPs affect height. Eventually, recombination would nominally create individual sets of genetic data containing all 262,144 possible combinations of the 18 SNPs. If the individual mutations each had a similar effect on height the range of the recombined effects would be 18 times greater than that of any single mutation. If the effects of individual SNPs on height varied as we would expect, the total range of variation between individuals would still be at least twice that produced by any one mutation.

We tend to think of mutation as something that happens *to* an organism and that the organism is helpless to alter or adjust the mutation process. Mutations occur regardless of the design of the organism. However, as we have seen in this chapter, the design of organisms grossly affects the mutational process. Some mutations are very rare. Some occur relatively frequently as a direct result of the design of the inheritance systems (e.g. variable number repeat sequences). Some occur even more frequently (e.g. mutational changes in white blood cell genetics associated with immunity). Pattern sensitivity and features like transposons affect the probability of and genetic location of mutations. Mutations are managed and processed by mechanisms that are part of the organism's inheritance system. In effect, mutations are merely the feedstock to a complex system that processes mutations. These observations are very important regarding the issue of evolvability to be discussed.

Biological Plans and Schedules

Any major project involving the construction of anything, say a house, involves plans and schedules. The *plan* (in this context) describes the physical locations of various components. There is a window here and a door there. Plans normally do not get too detailed. They do not specify the position of every brick but merely specify that a wall of certain dimensions is to be built of a certain type of bricks. The *schedule* specifies the *time sequence* in which the components will be installed. Some tasks can be performed in parallel. Other tasks can only be performed following the performance of some other portion or portions of the work. We *cannot* install the roof until the underlying structure is installed. We *must* install the roof before installing materials that would be damaged by rain. More complex portions of the work usually take longer to complete and involve more of these sequential tasks. It is usually beneficial to optimize the schedule to result in finishing the project as rapidly as possible to save time and therefore money.

The growth of a biological organism involves the same kind of processes except that the plan and schedule are genetically transmitted such that an organism constructs itself. Similarly to house construction, there is an obvious competitive benefit to be gained from more rapid development to maturity. In addition, in organism growth, complex structures such as eyes and brain tend to start development earlier. Finally, the genetic plan provides more detail for complex structures. Presumably, much more genetic code is involved in the specification of eyes and brain than large, simple, repetitive structures such as the *gluteus maximus*. As an example, the characteristics of major blood vessels are specified genetically. Everybody has an aorta. However, small blood vessels are not individually specified in the genetic plan but instead grow on an as needed basis.

Although different cells contain the same genetic instructions, different genes are activated in the growth and subsequent life of different cells. This accounts for their physical and functional differences. One mechanism whereby the different activations are implemented is a framework of chemical signals. As an organism grows, cells produce an expanding array of chemical signals that affect activation of genes in subsequent cells that then form different structures and systems. Such signals can either enhance or inhibit gene expression.

Enrico Coen, in his book *The Art of Genes[12]*, describes how the structure of a fruit fly is determined by this process. In humans and higher animals, at least one additional process, the progressive inactivation of certain genes as stem cells differentiate, also helps determine which genes are activated in a particular cell. The mechanics of this differential inactivation are not yet well understood.

In this connection, we tend to think of chemical signals such as hormones that circulate throughout the body. However, there are different levels of regionality. Depending on their solubility, diffusion characteristics, and other attributes, signals can be very local. Coen describes experiments in which transferring material from one part of a fly egg cell to another or from one part of an embryo to another resulted in the development of two-headed embryos and other structural abnormalities depending on the circumstances. Many signals are internal to individual cells and are even local to particular locations within a cell. A main purpose of the membrane surrounding a cell and a second membrane surrounding the cell nucleus is to help segregate signals by blocking some signals while allowing others to pass. The existence of a class of signals that are confined to cells allows processes that are controlled by those signals to proceed independently and simultaneously ("in parallel") in trillions of cells. Signals with longer ranges allow processes in groups of cells (tissues or organs) to be coordinated. Yet longer range signals allow signals from a gland or tissue to control processes in remotely located tissues.

There is communication between all the regionality levels. Signals can also be translated. A signal arriving at a cell can be received by a receptor which then produces another internal signal.

Even less well understood are the mechanics of biological scheduling. It is clear that development of complex organisms involves complex scheduling functions and that optimization of the schedule has an evolutionary benefit. Even at the cell level, there are many processes that must occur in a particular order. Chemical signals such as hormones and RNAs clearly play a role in scheduling of animal growth. The completion of a task often results in production of a chemical signal that activates genes to begin the next task and inhibits the genes that performed the completed function. Presumably, the scheduling aspects of an organism's development evolved in parallel with its structural and behavioral aspects.

The need for gene activation and deactivation and more specifically for a scheduling system to handle sequential biological activities has implications for aging theory. Apparently, not only must genes exist to perform some function, but some program for sequencing and programming those genes, (probably depending on yet other genes) must also exist. This would appear to be a major problem for traditional theories of aging which depend on the idea that random mutations exist which only cause a problem in older individuals or that genes can exist that have beneficial properties in youth but are adverse in older animals. (See Chapter 3.)

Chickens and Eggs

It is apparent from the discussion of genetic mechanisms that many issues along the lines of "which came first, chicken or egg" apply. It does not appear to make any difference what three letter sequence "codes for" a particular amino acid but the mechanism that reads the genetic code and assembles the amino acids must have the same rules that were used to write the code. For the system to work, the receiver of the message and the associated mechanism must have the same understanding as to the meaning of the various codons as the transmitter. While "Cat" in English means a furry house pet it could have just as well have been "Gato" or "Katz" as long as both the speaker and the listener had the same understanding. "CAT" in the genetic code calls for the amino acid histidine but is probably equally arbitrary. (Some organisms have been found that have slightly different correspondences between codon sequence and amino acid.)

The myriad chemical signals involved in gene regulation represent a similar situation. Each signal that is sent has no meaning unless receivers for that particular signal exist. The receivers have no function unless signals are being sent. There are presumably a very large number of possible signals. Any signal could be generated in a very large number of possible different locations within an organism. There can similarly be a very large number of possible locations within an organism (known as *receptors*) that are sensitive to any particular signal. Evolution of new signal/receptor pairs must take an extremely long time and be a very incremental process. Thus, genetics further validates Darwin's idea that evolution occurs incrementally but also suggests very long-term evolutionary processes. Both the genetic code scheme and the signaling scheme would appear to be very basic structures upon which the rest of an organism's genetic design is constructed. In fact, it appears that highly related animals such as mammals possess essentially the same genes and signaling schemes. Gene "A" produces a signal that affects genes "D", "F", and "K." The differences between mammals appear to be the result of minor differences in genes which mainly cause differences in degree. Gene "K" still responds to a signal from gene "A" but it takes more or less of the signal to have the same effect thus (for example) causing a particular mouse bone to be relatively longer or shorter than the equivalent human bone.

As another example, insulin serves the same function in all the mammals. Bovine insulin is so similar to human insulin (3 of 51 amino acids are different) that humans were able to use bovine insulin to control diabetes between 1923 and 1983. Human insulin is now manufactured using recombinant DNA in genetically engineered bacteria.

Evolutionary Genetic Processes

We can see from the foregoing that there are *at least* five separate processes involved in the evolutionary modification of the genetic code of more complex organisms.

The first and most rapid process is the recombination of the variable elements (i.e. single nucleotide polymorphisms) of the code through meiosis and crossover. Every individual animal represents a different combination of the variable elements. We can observe recombination by looking at a single generation.

In the second process, natural selection or selective breeding increases or decreases the population density of specific variations. Eventually a particular variable element allele could be eliminated from the population or become universal. However, natural selection or selective breeding *cannot* alter the perhaps 99+ percent of the genetic code that does not vary between individuals. Natural selection or selective breeding also *cannot* create variable code elements that do not already exist. We can easily observe the effects of this second process in breeding experiments or in observations of domesticated species.

In the third process, errors in copying or storing code (mutations) introduce new variable code elements in existing genes (e.g. single nucleotide substitutions). Presumably, because of the digital nature of the genetic system, most such new variations are sufficiently deleterious as to result in their being fairly immediately selected out. Occasionally, some have fitness effects that are initially sufficiently neutral that they avoid being selected out and eventually spread in the population to become polymorphisms which can participate in the second process. An error that causes a non-duplicated gene to have an entirely different protein product would appear to be unlikely to propagate even if it resulted in a potentially useful product because the original function of the gene and its benefit would be lost. A simple mutation such as a single letter substitution cannot alter a gene that does not already exist and therefore is limited in the scope of the changes it could cause in the design or behavior of an organism. In order to increase in complexity, an organism would presumably need to have *more* genes, not just changes to existing genes.

In the fourth process, entirely new genes are created by means of copying errors in which entire genes or partial genes are duplicated. The third process could then differentially modify the two genes such that they produce different proteins and thereby have more substantially different functions. In this way, additional genetic functions can be produced.

In the fifth process, genes are moved, or *transposed*, to different positions in the genome. Although such transposition does not affect the gene's function it does, because of the genetic distance principal, alter the inheritance patterns of genes and thereby alter evolution.

Speciation appears to be more dependent on organization of the genome than on content. It is clear that the mechanics of sexual reproduction including chromosome pairing, meiosis, and gene crossover depend heavily on a very high degree of similarity in the genetic organization (such as the number of chromosomes and order of genes on chromosomes) of the parents. Wild animals are observed that are nearly identical but nevertheless belong to different species. Domestic animals (e.g. dogs) are observed to be drastically different but belong to the same species. Speciation has a dramatic effect on the process of evolution because it prevents transmission of genetic characteristics to coexisting species.

Every species presumably inherited the vast majority of its genes from its ancestor species, some from very distant ancestors. Most of the genetic code therefore has a longer lifetime than the lifetime of any individual species.

Here are some examples of the potential complex interactions in these processes that are disclosed by our current fragmentary understanding of genetics.

Although alu elements have no known biological function, they could have a significant effect on the evolution of genetic code. The presence of alu elements in a particular region of genetic code increases the chance for subsequent duplications or deletions of code sequences (during meiosis because of the unequal crossover mechanism) in that region and therefore affects the fourth process.

Although introns in gene coding regions have no known biological function, the presence of introns could also affect the probability of and process of duplication. An alu or *within* an intron could affect the probability that *part* of a coding region might be duplicated or deleted, thus affecting the fourth process.

In addition to alu elements there are many other repeat patterns that could have similar effects on evolution of the genetic code.

We have to believe that the survival value of most variations is dependent on other variations. For example, a larger eyeball might be beneficial but only if accompanied by a larger eye socket. Now presumably there are genes that affect the size of the whole animal, there are genes that affect the size of the head relative to the rest of the animal, and there are genes that affect the size of the eyeball relative to the rest of the head. Similarly, physical characteristics of organisms must be matched by appropriate behaviors and behaviors must be matched by appropriate neurological systems. It is clear that evolution would be assisted if inheritance of certain genes was associated with inheritance of certain other genes such that, for example, eye size tended to be associated with eye socket size. We know that the degree of such association depends on the relative physical location of the genes in the genetic code of a chromosome. It "boggles the mind" to contemplate how long it could take to achieve these kinds of associations through the process of gene copying and transposition. Presumably, many such associations as well as their underlying genes are inherited from ancestor species and have long lifetimes relative to a "species lifetime."

Because mutations in junk DNA presumably have no biological effect, such mutations can propagate easily through a population. The presence of these mutations could then affect the probability of subsequent mutations through various forms of "pattern sensitivity" and processes such as described above and thereby have significant long-term effect on the evolution of that species' genetic code.

It is estimated that only about 1 percent of human DNA is in the form of gene exons. If we include in "functional DNA" all the regulatory regions, leading and trailing patterns in introns that cause them to be introns, the patterns in telomeres and centromeres that cause them to function, and all the other DNA that seems to have some fitness effect, the total functional DNA is probably less than ten percent of the total genome.

Suppose we were to rearrange the genome of a mouse. We could take the same mouse genes (excepting the sex chromosomes) and place them in different positions on different chromosomes. We could even equip our new mouse with a different number of chromosomes. Because the new mouse has the same genes, the mice in a population of new mice should be physically indistinguishable from old mice. They should have the same fitness as old mice. The only difference is the order in which the genes are sequenced in their chromosomes.

We can make some more changes. We could add introns to some genes and delete introns in other genes. We can change the specific internal sequences of introns. We can add, or delete, or change the content, of other junk DNA. As far as is now known, none of these changes would affect the appearance, behavior, or fitness of the new mice relative to the old mice.

However, it is clear that we might be drastically altering the ability of the new mouse to adapt and evolve. Since genes are in a different order, genes that formerly tended to be inherited as a group because of short genetic distance would now be independent. Other genes formerly on different chromosomes could now be in clusters. Protein "modules" formerly available because of intron structure would no longer be available. Although the new mouse is physically and functionally identical to the old mouse, the mechanics of evolution available to the new mouse would appear to be significantly different. Because of the major differences in genome organization, the new mice would be unable to interbreed with old mice. The new mice would be members of a different, though physically and behaviorally identical, species.

Because of the large differences in genome organization between similar species (e.g. mammals) it is clear that the *organization* of the genome evolves. This evolution must necessarily be extremely "incremental" (more "tiny steps") in order to maintain the ability of individual members to interbreed.

Speciation may be more important to the process of evolution than widely thought. For example, a mutation in junk DNA can propagate relatively freely through a species population but not to coexisting species. Because the mutation had no biological effect, it would not be selected in or out by natural selection. However, such a mutation could alter the probability that *other* specific mutations could occur in that species thus potentially blocking avenues of evolution that were still open to similar coexisting species. An individual species could become extinct, not because it had any less fitness, but because its genetic code had been, in effect, poisoned, regarding its ability to adapt as well as competing species. Conversely, such a mutation could open up avenues of code evolution not available to coexisting species by increasing the probability of certain mutations. This in turn suggests that survival of *species* could be more important relative to individual survival in the overall process of evolution and improves the case for species-level group selection.

Imagine that a builder sends an email to a subordinate that reads: "Build a red brick wall 910 cm wide, 220 cm high and 20 cm deep." The builder uses the Eudora email software. The subordinate can use any email software *as long as it understands the protocols and formats used by the sender.* The builder wrote the message in English but it could have been Spanish *as long as the receiver understood Spanish.* The builder could have hand-written and hand-carried a note with this message *as long as the receiver was able to read his handwriting.* Nobody would deny that the complex languages, protocols, and formats used to convey the message represent complex designs and yet they do not have any effect on the functionality of the wall. The wall is physically the same, performs exactly the same function, would last as long, and is exactly as "fit" regardless of the method used to convey the message. The benefit of communications improvements is extremely diffuse, long-term, and aids all builders.

As we have seen from this chapter, the complex systems that provide for inter-generational communications of organism design characteristics have exactly the same kinds of considerations. Fitness does not appear to provide a mechanism for driving the evolution of these complex systems. *The media is not the message.* It is apparent that the communications methods have only an extremely long-term effect. This sort of logic provides important support for the existence of long-term evolutionary processes and supports evolvability theory to be discussed.

This entire scenario appears to be incompatible with classical Darwinism as follows. Darwin's theory holds that mutations that are beneficial to an organism increase its fitness and mutations that are adverse decrease its fitness. Mutations with positive or minor fitness impact can eventually become widely distributed in a species gene pool. Period.

In the above discussion, we have identified a whole family of different types of mutational changes which have no immediate fitness effect but which plausibly benefit or detract from the ability of the organism to *subsequently* adapt through evolution by reducing or increasing the probability that certain types of subsequent mutation can occur and also altering the probability that certain genes will be inherited in conjunction with specific other genes. In effect, these mutations affect the *future* of the organism in terms of the descendent species it might produce or the "evolution of its species." Although such mutations, either beneficial or adverse, could, (since they are fitness neutral), spread through the population of a species and could be transmitted to any descendent species, they cannot spread to co-existing species. This suggests that "survival of the species", that is, species that produce descendent species or which "evolve", as opposed to becoming static or extinct could play a much more important role relative to "survival of the fittest individual" than contemplated by classical Darwinism. In addition, speciation, per se, is substantially the result of such non-fitness mutational changes and itself obviously plays an important role in evolution.

"Evolution of the genome" tends to support modifications and adjustments to classical Darwinism such as the selfish gene theory and evolvability theory.

The purpose of this chapter has been to illustrate the complexity that has appeared as we discovered more about the mechanisms whereby evolution of genetic codes actually occurs. Darwin's analog world is very simple when compared to the digital reality. Breeding experiments and heredity studies are generally confined to exploring recombination and natural selection. Variation and natural selection are essentially the easily observable "tip of the iceberg" regarding the mechanics of genetic code evolution. The time scales of these different processes differ enormously. The selection of a trait that was represented in variations could take a relatively few generations. Other traits, produced or affected by non-variable parts of the genome code could be conserved for millions or billions of years. Details of the mechanics of the third, fourth, and fifth processes could explain why Darwin's theory does not work for aging and other troublesome animal characteristics. Specifically, it appears that evolution could involve much longer times and more complex processes than contemplated by orthodox Darwinism and that therefore the importance of "individual" fitness could be less than considered by Darwin. At the same time, knowledge of these complex processes supports Darwin's determination that sudden massive mutations were unlikely to have a significant role in evolution.

Genetics science is largely "applied" science having many applications with major practical importance in disease diagnosis and treatment, forensics, pharmaceutical manufacture, and development of genetically engineered plants and animals. Although many genetics discoveries have implications for the "pure" or "academic" science of evolutionary mechanics, they tend to be a very secondary concern.

6. Discoveries Affecting Aging Theory

This chapter is intended to summarize discoveries, observations, oddities, and other information that particularly offer insight into the aging process or underlying evolutionary mechanics theory. Some of the discoveries and developments post-date the traditional aging theories and affect their credibility.

Senescence of Salmon

Salmon are interesting in that they display one of the most spectacularly aggressive aging mechanisms, essentially biological suicide, and have other characteristics that appear to be incompatible with popular theories of aging and even Darwin's mechanics.

Salmon are hatched in fresh water streams. The young fish eventually (after as long as one year) migrate to the ocean and are thought to range over long distances. Mature salmon return to their stream during mating season (spring) and swim upstream as far as 500 miles (800 Km) to their spawning area. The salmon are able to adjust between fresh and salt water in both directions. Upstream travel, including fish ladders and rapids presumably requires fish to be in excellent physical condition.

Following spawning, the adult salmon die of old age, usually within a week. Salmon exhibit generalized, multi-tissue deterioration during this aging process.

The pink salmon mate and die after two years. Other varieties have longer lifespans but are semelparous and only mate once. Yet other varieties can survive migration and mating one or more times before dying. The aging mechanism onset is apparently triggered by reproductive activity or by whatever triggers reproductive activity. Aging salmon display physiological changes such as a "hump back" appearance and changes to the shapes of their jaws.

-Different varieties of salmon display very great variation in lifespan characteristics between very similar organisms.

-The salmon are examples of "acute senescence" as opposed to gradual degradation seen in many animals.

-Biological suicide in the salmon is highly structured and associated directly with reproduction, a specific season, and migration, rather than calendar age.

-Wild salmon actually do "die of old age."

These characteristics of salmon clearly are not compatible with gradual accumulation of damage. They appear to be generally incompatible with theories that involve inadequacy of maintenance or accumulation of mutations.

The salmon also appear to represent a discrepancy with orthodox mechanics. It is difficult to explain the behavior of the salmon as anything other than an example of programmed death. Programmed death is not necessarily incompatible with Darwin's theory if a valid tradeoff exists. In this case, the tradeoff would have to be between the disadvantage of dying after spawning and not producing subsequent descendants, and some benefit to the current immediate descendants of the *individual* dying salmon.

Some scientists theorize that the decomposing bodies of dead salmon in streams nurture young salmon immediately after they hatch and that therefore the accelerated aging produces benefit relative to a situation in which the adult salmon survive the stream and die subsequently in the ocean in a more conventional life cycle. If the young salmon were direct descendants of the adults, death of the adults could produce an individual benefit. However, there are issues. The fish that died after spawning could not be sure that its personal descendants would benefit from the nutritional value of its corpse to a greater degree than salmon that did not die after spawning. Another salmon, not programmed to die, could be equally likely to benefit. Since the second fish could go on to have subsequent descendants, it would be more fit and would be expected to propagate its design better than the suicidal variety. This is an example of what evolutionary biologists refer to as the "cheater problem." Many proponents of non-programmed mammal aging accept that the salmon are examples of "special cases" in which programmed death produces an individual benefit.

The salmon and the many other species displaying programmed death are more compatible with adjustments to Darwin's theory that deemphasize individual fitness as described in Chapter 7.

Elephant Teeth

Humans have two sets of teeth. In prehistoric times, loss of many teeth in the second set presumably led to weakness and increased mortality and therefore clearly was a fitness factor.

If a human loses a tooth from the second set, it is not replaced, regardless of the age at which the loss occurs. Apparently, humans are genetically programmed to only have two sets of teeth. Failure to grow more teeth is not a function of age.

Elephants, which as herbivores eat almost continuously, have 6 sets of teeth, each of which gradually wear out and is replaced by new teeth. When the last set wears out the elephant starves. Some refer to this as an example of "mechanical senescence."

Elephants have few predators. Some observers report that death from tooth loss, effectively "death of old age", is not uncommon in wild elephants. If true, this is a problem for the mutation accumulation theory which proposes that aging has no fitness effect.

Why don't humans and elephants have more sets of teeth? If nature can have two sets or six sets then there is clearly no fundamental limitation that would prevent additional sets. The absence of additional sets of teeth appears to have results functionally similar to aging, while at the same time appearing to be a substantially different type of phenomenon from the sort of generic, time-sequential, deterioration typical of other aging characteristics. The situation with teeth strongly suggests programmed death.

Fruit Fly Longevity

Fruit flies (*Drosiphila melanogaster*) are a favorite of genetics researchers because they have a relatively short lifespan and are easy to breed (as we all know too well). Experiments have been performed to see if selective breeding could produce longer-lived fruit flies. The experiment was arranged such that the longest-lived flies of a generation were mated to produce more flies which were also selected for longevity and so on. After approximately 29 generations, the lifespans of the flies were increased 50 percent[13].

However, other investigators question on technical grounds whether any increased longevity has actually been demonstrated[14].

The increase in lifespan, if present, appears to be a survival advantage, an improvement in the fly's ability to compete, so what would happen if we released some of these flies into the wild population? Would the wild flies eventually all have the longer lifespan? Most biologists would say "no." The selectively bred flies are essentially domesticated. In selectively breeding for longevity, we presumably inadvertently changed other characteristics of the flies that resulted in survival disadvantages. Although the new flies have no obvious disadvantages such as shorter wings or reduced ability to breed, more subtle disadvantages are bound to be present as a consequence of the fact that selective breeding can only alter the relatively few design characteristics that vary between individuals in the population. The new flies would therefore be at a fitness disadvantage relative to the existing wild population and would die out.

Mutation Experiments

Experiments have been performed in which mutations were artificially induced. For example, fruit flies can be exposed to radiation, vastly increasing the rate at which random mutations occur. When fly embryos are examined under a microscope, all sorts of gross abnormalities such as more or less than the normal number of major parts (head, wings, legs), or obvious abnormalities to major parts can be observed. These gross and adverse changes correspond to Darwin's "monstrosities." *Adverse* mutation is easily demonstrated.

Experimentally demonstrating a random *beneficial* mutation in a complex organism (even as complex as a fruit fly) would be extremely difficult. Such a mutation would be very small and therefore very difficult to distinguish from the non-mutant organism. Beneficial mutations would be extremely infrequent. Demonstrating that a mutation was actually "beneficial" would involve introducing the mutant strain into a non-mutant, presumably wild, population and observing how well they were able to compete under wild conditions. Independently repeating such an experiment would clearly be essentially "undoable." Exploration of beneficial mutation in complex organisms, like many aspects of Darwin's theory, is not a suitable subject for experimentation.

Canine Longevity

Larger wild animals tend to have both longer times to develop to sexual maturity and longer lifespans. However, larger dog breeds while taking longer to develop tend to have *shorter* lifespans than smaller breeds. Some smaller breeds are said to have maximum lifespans of as much as twice that of larger breeds.

We presume that essentially all the *differences* between dog breeds are the result of selective breeding rather than natural selection. All the dogs have the same chromosomes and presumably the same genes. The differences between breeds are genetically expressed as variations in the genes (e.g. single nucleotide polymorphisms). The lifespan differences are accidental. Breeders over the centuries were presumably not trying to develop a dog with a longer or shorter lifespan but were trying to develop other qualities. If breeders had been trying for centuries to breed longer and shorter lived dogs, the lifespan differences might be similar in magnitude to the differences in size or other physical parameter.

The canine longevity differences are a demonstration that lifespan is a property that can be affected by selective breeding and therefore also presumably by natural selection. The canine case also demonstrates that there is not a fixed relationship between development time and aging.

Insect Life Cycles

The cicada (*Cicadidae homoptera*), commonly known as the 17-year locust, lives in the ground as a larva for 17 years, then emerges from the ground as a winged adult, mates, lays eggs, and dies. The adults generally live only a week. The cicada is one of the longest living insects and an example of an extremely precisely programmed life cycle in which one stage is dramatically longer than the others. A major curiosity is the fact that most of the insects in a "brood" emerge from the ground within 24 hours of each other. This is thought to be a survival characteristic in that birds and other predators would tend to be overwhelmed by the simultaneous appearance of the insects. Mating is also obviously facilitated. Both of these characteristics convey individual benefit and are therefore compatible with orthodox Darwinism. Either the emerged insects can signal the un-emerged insects (possibly by the loud noises they generate) and/or they possess very accurate biological clocks.

Most of the cicadas have, within about 0.1 percent, the same lifespan. A traditional grandfather clock only has an accuracy of about 0.2 percent. Cicadas therefore represent one of the most definitive examples of "programmed lifespan."

The mayfly has a life cycle one year long. The adult form does not have a functioning digestive system and therefore cannot eat. It only lives for a few hours or days. This certainly would appear to be a case of programmed death.

Insects also have dramatically different functional designs at different stages in their lives. This demonstrates that the same organism can evolve conflicting design properties at different ages. This adds to the support for programmed aging in mammals by responding to critics that ask how a mammal could purposely evolve both strength and weakness, good eyesight and poor eyesight, etc.

Spider Suicide

Some female spiders eat their mates after mating. Some observers report that the males appear to commit suicide by jumping into the female's jaws or otherwise cooperating in their demise. Praying mantis females also eat their mates. This could represent individual benefit since adult mantids do not survive winter.

Plant Suicide -- Bamboo

Bamboo displays another oddity in the world of aging theorists. Bamboo propagates by extending lateral roots or runners. New shoots extending from the runners eventually develop into mature plants. The plants developing from the runners are essentially clones of the original plant.

Approximately every 100 years (some varieties have shorter cycles), a stand of bamboo produces flowers, produces seed, and dies. This is one of a number of examples of a relationship between lifespan and reproduction and is also cited as an example of essentially suicidal semelparous behavior or "acute senescence" as opposed to "gradual senescence."

Notice that the individual plants in the stand of bamboo, complete with stems and roots, have different ages. Presumably, the plants toward the edge of the stand are progressively younger. Nevertheless, all the plants die at the same time. The actual age of the individual offshoot plants does not appear to be a factor in when they die. Death is triggered by reproduction. This is an illustration that aging is not necessarily associated with growth and also one of the clearer examples of programmed death.

Some theorize that the bamboo dies in order to force at least periodic sexual reproduction and avoid a situation in which reproduction is dominated by cloning.

Many plants exhibit similar semelparous behavior in which death is triggered by sexual reproduction.

Semelparous Mammals

A species that sexually reproduces only once is known as *semelparous*. Usually the organism dies more or less immediately after reproducing. This behavior is observed in many plants and some animals including octopus, and some salmon. Most mammals display gradual aging. However, at least one mammal, the male *Antechinus stuartii,* is semelparous[15]. This small Australian marsupial mouse has an annual mating season. Males survive until the following mating season, mate, and die. Females typically live only long enough to wean their young but sometimes survive several years.

Some biologists, even believers in traditional evolutionary mechanics theory, concede that semelparous species represent instances of programmed death, that is, death caused by an evolved mechanism whose specific purpose is to limit lifespan. Traditional biologists contend that the semelparous species *needed* to limit their lifespans for reasons that do not apply to the majority of (gradually aging) mammals and therefore evolved suicide mechanisms. The suicidal behavior is said to result in individual benefit of some sort. However, the traditional explanations are typically weak and sometimes entirely circular repetitions of Darwin's "explanation": There must be some hidden individual benefit because traditional evolutionary mechanics theory says so.

Many traditional biologists alternately consider that semelparous organisms die of "exhaustion" related to their reproductive activity and therefore are not examples of suicidal behavior. They die because of a tradeoff between the advantage of living longer (small or negligible according to two popular value-of-life theories) and the competitive advantage of more vigorous reproductive effort. They have difficulty explaining why similar organisms perform the same reproductive tasks without dying of "exhaustion."

If lifespan control is a necessary biological function we would expect organisms to have evolved different ways of accomplishing that function. Animals all need mobility but have evolved myriad ways (flying, crawling, digging, etc.) to accomplish mobility.

Non-Aging Species – Negligible Senescence

Negligible senescence is defined as a situation in which an organism does not exhibit measurable decline in fitness with age. It does not show decline of survival parameters such as strength, agility, or sensory acuity with age and also does not exhibit reduction in reproductive capacity with age. Mortality rates in the wild for a negligibly senescent species are constant (after developmental maturity) and might actually decline as described below.

As shown in chapter 1, some sturgeon, some rockfish[16], and some turtles have extremely long maximum recorded lifespans (in excess of 140 years). In some animals, age of a captured specimen can be determined from annual rings, similar to tree rings, which form in scales or certain bones. These animals grow slowly and take a long time to reach sexual maturity so their long lives fit some traditional theories as well as new theories of aging.

However, these animals apparently do not age. Scientists disagree on whether they do not age or age very slowly but functionally these animals having negligible senescence do not appear to deteriorate as they get older. An animal such as a rockfish, caught in the wild at age 140, presumably had not been significantly weakened by age. If it had been, it would have succumbed to predators. These animals have little or no observed increase in mortality rate with age. Older specimens are not more susceptible to disease. Older animals do not display a reduction in reproductive capacity or vigor. In some cases, claims have been made that reproductive capacity (as might be indicated by number of eggs produced) actually increases with age. These animals do not appear to display any decline in strength or agility with age.

Non-aging animals still die from predator attack, warfare, accident, disease, inability to obtain food, and environmental conditions but the probability of such death is not a function of age beyond full maturity. Very old individuals are consequently rare. Infant mortality and death of immature specimens are probably similar for aging and non-aging species.

Some claims of *negative senescence* have been made suggesting that death rates actually decline with age. This would be expected in a non-aging species because of the effects of acquired fitness advantage such as conferred by intelligence or immunity.

Figure 18 Rougheye Rockfish

Since there are only a few non-aging species amid many fairly similar aging species, the non-aging animals must be descendants of aging species. They appear to have *lost* the ability to age. Non-aging animals whose reproductive capacity does not decline obviously represent a major problem for the disposable soma theory.

Non-aging animals present an enormous research opportunity. We should be able to identify genes and associated processes and mechanisms that are unique to the aging animals or unique to non-aging animals.

The oldest known non-clonal living organism is a tree called "Methuselah." This tree, a Great Basin Bristlecone Pine (*Pinus longaeva*) in the White Mountains of California has been determined to be 4839 years old (in 2007) by annual ring count observed by a boring.

The oldest known clonal colony is a quaking aspen tree (*Populus tremuloides*) known as "Pando" and located in Utah. This colony shares a single root system serving 47,000 stems, occupies 107 acres, and is thought to be as much as 80,000 years old. Individual stems are only thought to be perhaps 150 years old. Theorists suggest that the reason Pando does not die is that it does not flower and seed – if it did it would die like the bamboo. They suggest that it does not flower and seed because some condition (such as ice-age environmental conditions) that formerly triggered sexual reproduction and subsequent suicide in this species no longer exists.

Naked Mole Rat (NMR)

The naked mole rat (*Hetrocephalus glaber*) is a mouse-size burrowing rodent native to East Africa. Mole rats are the only mammals that exhibit *eusociality* and have a social structure similar to colony insects. In a colony (typically 75 individuals) there is a queen, several fertile males, and sterile workers. The lifespan of the longest-lived NMR has been measured at about 28 years, longer than any other rodent and much longer than similarly sized mice (~3 yr). NMRs are considered by some researchers[17] to be negligibly senescent in that they do not exhibit fitness decline or increased mortality with age like gradually aging mammals. In particular, reproductive capacity does not decline with age. Unlike other mammals including some shorter-lived rodents NMRs do not naturally develop cancer, obviously interesting to cancer researchers. The NMRs have attracted considerable attention as models for the study of aging mechanisms.

Figure 19 Naked Mole Rat

NMRs in the wild have an annual mating season. However, in captivity, they are not inhibited by any seasonal limitation and can reproduce up to four times per year. Apparently, planetary cues that inhibit reproduction in the wild are absent in the laboratory setting.

Aging Genes and Lifespan Regulation Discoveries

Recently, "aging genes" have been discovered in the roundworm, fruit fly, and mouse. Disabling these genes has resulted in lifespan increases of between 30 and 600 percent. These species are among the most heavily genetically mapped and studied of all species so such genes are probably widespread.

For example, Cynthia Kenyon[18], a researcher at the University of California in San Francisco has discovered a gene in the roundworm, which, when altered, doubles lifespan:

> "The nematode *C. elegans* ages rapidly, living just over two weeks. We have found that a gene called daf-2, which encodes a protein resembling the insulin/IGF-1 receptor, regulates the rate of aging in this animal. Changing the activity of this gene can double the lifespan of the worm, keeping it healthy and youthful much longer than normal. Our findings indicate that DAF-2 is part of a multi-step signaling cascade, and that it acts by regulating the activity of a transcription factor called daf-16. This hormone system responds to signals from the reproductive system that regulate aging, and to signals from the environment. Together our findings indicate that the aging process in this animal is surprising plastic, and that DAF-2 and DAF-16 act as part of a "central processing system" to integrate signals with many different origins that together define the aging rate of the animal."

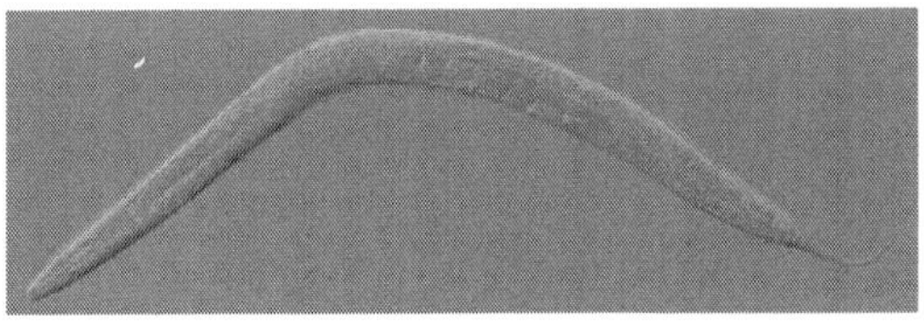
Figure 20 *C. elegans*

Aging in the relatively simple roundworm seems to be a very complex process involving "signals with many different origins" and suggesting that aging is an adaptation rather than the result of random events. Kenyon subsequently reported experiments in which treatment of multiple genes resulted in lifespan extension by a factor of six.

So far, no one has discovered a function for these genes other than causing aging. Organisms lacking the gene seem to be normal except for having lost the ability to age as rapidly as unaltered specimens. Of course, proving that the genes did not have some individual benefit would be very difficult. In fact, it is unlikely that such a gene has no other phenotypic effect. However, the other effects may well be minor.

Since the organisms still age at a slower rate, additional genes and mechanisms must be involved in aging.

Goldsmith

Aging genes that have no other function represent a major problem for the antagonistic pleiotropy theory and disposable soma theory. Aging genes also represent a problem for the mutation accumulation theory in that creation of an entire functioning gene that is adverse to fitness by means of random mutations seems implausible.

Aging genes that have no other function are incompatible with the traditional aging theories and orthodox evolutionary mechanics theory.

Apfeld and Kenyon report[19] finding lifespan regulation by detection of external signals in the roundworm. This is a direct discovery of non-genetic adaptation applied to lifespan adjustment in adapting to external conditions:

"Caenorhabditis elegans senses environmental signals through ciliated sensory neurons located primarily in sensory organs in the head and tail. Cilia function as sensory receptors, and mutants with defective sensory cilia have impaired sensory perception. Cilia are membrane-bound microtubule-based structures and in C. elegans are only found at the dendritic endings of sensory neurons. Here we show that mutations that cause defects in sensory cilia or their support cells, or in sensory signal transduction, extend lifespan. Our findings imply that sensory perception regulates the lifespan of this animal, and suggest that in nature, its lifespan may be regulated by environmental cues."

Wolkow, et al[20] have similarly discovered internal signaling controlling a lifespan regulation mechanism in the roundworm:

"An insulinlike signaling pathway controls Caenorhabditis elegans aging, metabolism, and development. Mutations in the daf-2 insulin receptor-like gene or the downstream age-1 phosphoinositide 3-kinase gene extend adult life-span by two- to threefold. To identify tissues where this pathway regulates aging and metabolism, we restored daf-2 pathway signaling to only neurons, muscle, or intestine. Insulinlike signaling in neurons alone was sufficient to specify wild-type life-span, but muscle or intestinal signaling was not. However, restoring daf-2 pathway signaling to muscle rescued metabolic defects, thus decoupling regulation of life-span and metabolism. *These findings point to the nervous system as a central regulator of animal longevity.*" (Emphasis added.)

Progeria and Werner's Syndrome

There is a rare human disease, *Hutchinson-Guilford progeria syndrome*, in which aging-like conditions are greatly accelerated (about seven times normal rate) to the point where individuals usually die by age 14. Hutchinson-Guilford syndrome has an incidence of about one in four million births and is thought to be caused by a dominant single gene mutation in each victim. Hutchinson-Guilford progeria patients display accelerated heart disease, baldness, gray hair, and other symptoms associated with aging.

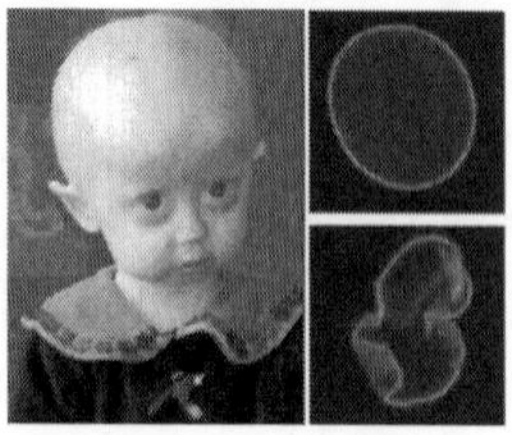

Figure 21 Progeria Victim

A variant, *Werner's syndrome*, also a result of a single gene mutation, causes somewhat less accelerated aging resulting in victims dying usually by age 50. Patients with Werner's syndrome have many age-related conditions and diseases including atherosclerosis and other heart disease, baldness, gray hair, joint disease, skin conditions, cataracts, osteoporosis, some cancers, and diabetes, in addition to other conditions such as dwarfism and other malformations. Cells of Werner's patients display characteristics associated with aging such as reduced division potential.

Research on progeria and Werner's syndrome victims could lead to important insights into the operation of the aging mechanism.

Progeria and Werner's syndrome also provide an important clue that aging is centrally controlled such that a single genetic malfunction can invoke all the aging symptoms that have been observed. Specifically, a theory such as the antagonistic pleiotropy theory (or the passive maintenance theory to be described) that holds that aging is a result of many independent genes appears to be incompatible with these conditions. Many different genes in many different tissues are no doubt involved in aging but they seem to be centrally *controlled* or *regulated* in such a way as to allow a single genetic error to result in accelerated aging. Progeria and Werner's syndrome suggest that aging is a deliberate evolved mechanism of nature and not an accident.

Caloric Restriction

Researchers (Weindruck, et al and others) have observed[21] that caloric restriction (CR), the feeding of a nutritional but reduced diet, in many animals including primates causes a dramatic increase in average and maximum lifespan (more than 50 percent) under lab conditions. As caloric intake was restricted to as little as 30% of the amount the animal would eat if given unrestricted access to food, lifespan increased almost in direct proportion to the reduction. Reproduction of animals under caloric restriction is reduced and, at lower feeding levels, eliminated.

An obvious thought is that bodily processes in the CR animals are generally slowed down by the reduction in food energy. The slow-down in metabolism could include a slowing of the aging process. CR could result in a case of partial "suspended animation." However, this does not appear to be the case. CR animals are actually *more* active than control animals.

The lifespan increase is accompanied by a delay in the appearance of features generally associated with aging such as increased susceptibility to diseases, weakness, etc. The animals stay "younger", more active, and healthier longer, not just live longer. CR can apparently reverse effects of aging even if applied only late in an animal's life. An especially interesting finding is that the CR animals appear to have lower rates of some cancers.

Researchers Weindruch (University of Wisconsin) and Spindler (University of California) say:

> "Caloric restriction is effective in diverse species. Most caloric restriction studies have been in rats or mice. However, caloric restriction also extends lifespan in single celled protozoans, rotifers, water fleas, fruit flies, spiders and fish. Restricted animals stay biologically 'younger longer.' Caloric restriction in mice and rats extends biologic youth and postpones or prevents most major diseases (cancers, kidney disease, cataracts, etc.). Accordingly, the caloric-restricted rodent provides a model to study aging with minimal distortion from diseases. About 90 percent of the 300 or so age-sensitive outcomes studied stay 'younger longer' in caloric restricted animals. For example, decreases in certain immune responses begin in normal mice at one year of age, but begin at two years of age in restricted mice."

The existence of the CR phenomenon provides an important tool for development of treatments: A method might be found for simulating or stimulating the anti-aging effect of CR without experiencing CR itself. A CR mimetic, 2DG, has already been found[22] which mimes the effect of CR in animal trials.

Gene chip studies are being performed comparing gene expression (see Genetics) in CR animals to control animals to find differences that might lead to identifying CR related genes and thereby, potentially, aging related genes. This could lead to identifying a reliable indicator (See the Indicator Problem) of aging so that effectiveness of potential anti-aging treatments could be rapidly evaluated. Currently researchers have to "wait for some rats to die" in order to have a generally accepted indication that an anti-aging treatment is effective. Human trials using this approach could take a very long time for each trial!

Could the relaxation in the aging mechanism be an adaptive evolved response to starvation? If starvation was the result of an event such as drought, a reasonable response might be a relaxation of the aging mechanism combined with a reduction in birthrate (also observed in starvation conditions). Maintaining an adult population would require fewer resources than producing and supporting young and would position the population for rapid reproduction once the event was ended. This line of thought is a logical consequence of the idea that organisms possess a lifespan control mechanism that is capable of non-genetic adjustment (regulation) in response to temporary or local conditions. As indicated earlier, such self-adjustment capability is common in evolved biological mechanisms. The CR observation therefore logically fits with an evolved lifespan regulation mechanism but does not fit with traditional non-programmed aging theories.

The CR observation appears to be particularly incompatible with the disposable soma theory that claims aging is due to resource limitations.

Stress and Aging

As we have seen in the previous section, restricting caloric intake causes a counterintuitive increase in lifespan. It is also well known that exercise improves health. In a spectacularly counterintuitive way, a variety of experiments have found that *many types of stress seem to increase lifespan*:

-Exposure to low dose radiation extends lifespan in rats and fruit flies.

-Regular exposure to electric shock extends life in mice.

-Periodic immersion in cold water results in longer rat lifespan.

-Low doses of chloroform have been observed to increase lifespan in dogs.

The nature of the stress effects provides further evidence that aging is not simply a result of accumulation of damage and generally presents a difficulty for non-programmed aging theories.

If aging is the result of a programmed lifespan control mechanism capable of self-adjustment, the stress response is a logical local or temporary adjustment to lifespan. If a population was under heavy predation, certainly its members would feel stress and fear associated with escaping from or fighting predators. Under such circumstances it would make logical sense to increase lifespans of organisms to compensate for population losses caused by the predators and therefore better maintain the population. In a population living in a relatively predator-free area, the stress and associated response would not exist. The same sort of stress response might be useful in case of severe environmental conditions. Degree of predation, availability of food, and stress due to environmental conditions are all conditions that would tend to vary in a local or temporary way and therefore subjects for non-genetic adjustment of lifespan. The optimum lifespan for an organism logically would vary in response to all of these conditions.

Octopus Suicide

The female octopus displays what appears to be a very explicit example of programmed death. The octopus, which normally only reproduces once stops feeding and dies shortly after reproducing. However, surgical removal of the optic glands prevents (Wodinsky, et al[23]) this result and the animal begins feeding again and survives for at least another breeding season. Apparently, the optical apparatus provides some hormonal signal to activate the programmed death mechanism.

Here we have a case of not only programmed death but actual programmed suicide or death resulting from a behavior. The animals die of starvation. They starve because they do not eat. They do not eat because they do not experience hunger. Hunger is controlled by hormones. Therefore, lifespan in this instance is controlled by a complex central mechanism.

Figure 22 Octopus

This characteristic of the octopus has two very interesting attributes. First, unlike the salmon, there does not appear to be any obvious potential orthodox Darwinian (individual) benefit to descendants resulting from death of the parent. Second, lifespan control involves a behavior and suggests that *other* behaviors might be significant in regulating lifespan.

The octopus lifespan control mechanism involves communication with the central nervous system in both directions: Sense organs are involved in detecting the circumstances leading to activating the suicidal behavior, and, the execution of the behavior involves inhibiting the organism's normal hunger response, a nervous system function. This is a textbook example of the sense/process/execution organization seen in non-genetic adaptation.

A traditional explanation for organisms that die after reproducing is that they die of "exhaustion" associated with reproduction. The idea is that they are the result of a tradeoff between more vigorous reproduction and a longer life, a tradeoff that is supported by traditional evolutionary mechanics. It seems obvious that this explanation cannot be applied to the octopus as the specimen successfully reproduces and then survives (even with surgery) if the suicide mechanism is inhibited.

Any discussion of animal suicide invariably results in someone mentioning lemmings. Some have observed lemmings apparently jumping off cliffs or into water. However, current scientific consensus is that lemmings do not commit suicide but rather die by accident (pushed off cliffs by other lemmings) or intentionally jump in a non-suicidal effort to cross an obstacle during mass migration.

Sex and Aging

There have long been reports that people having more sexual activity tend to live longer. (Some might suppose that this is related to the stress effect!)

In 1997, Smith, Frankel, and Yarnell published a paper in the British Medical Journal titled *Sex and death: are they related?[24]*. This 10-year study of 918 men from the area around Caerphilly, South Wales correlated "orgasmic frequency" with mortality. The study found that men (aged 45 – 59 at the start of the study) with a high (twice per week or more) frequency of orgasm had a mortality risk 50 percent lower than men reporting a low (less than monthly) frequency. Heart disease risk was especially beneficially affected.

Obviously a major issue here is determining whether sex causes good health or good health causes more sex. Healthier people would be expected to be more sexually active. The authors tried to compensate for health issues at the time sexual activity was reported but admit that proof of cause and effect is difficult. Similarly, an individual in whom the aging process was progressing more slowly, for whatever reason, could be expected to be more sexually active than other individuals of the same calendar age.

Orgasm is known to be associated with release of the hormone oxytocin. Synthetic oxytocin (pitocin) is used to induce labor.

The findings of this study conflict with much cultural and popular opinion to the effect that sex is debilitating. They also conflict with aging theories (especially the disposable soma theory) which hold that increased reproductive activity should be matched by decreases in lifespan.

Non-deteriorative Human Aging

We tend to think of aging in humans as being entirely physically deteriorative in nature. While this is largely the case, there are a few examples of apparently genetically programmed changes that are not deteriorative in nature and extend to advanced ages. Many of the externally observable physiological changes that occur with age are indeed plausibly the result of deterioration, such as the reduction in head hair observed in most people. However, in some areas, such as ear hair, growth actually increases with age. Changes of appearance between middle age and old age are to some extent the result of redistribution of tissue rather than deterioration. These observations support the idea that a biological clock exists that continues to program activities throughout life.

Similarity in Aging Symptoms

In gradually aging mammals, the symptoms of aging tend to be rather similar, even among species having dramatically different lifespans. People who are familiar with cats and dogs know that they typically die of the same age-related conditions (such as heart disease and cancer) that affect humans. They also typically suffer from cataracts, deafness and other sensory deficits, arthritis and other mobility deficits, and other typical symptoms of aging. As will be discussed, this observation fits programmed aging theories better than it fits non-programmed theories.

Programmed Cell Death

It has been known and generally accepted for some time that some cells are genetically programmed to die. Programmed cell death, known as *apoptosis* occurs during the lives of many plants and animals including humans. For example, some mature plants have leaves with holes. The holes are formed when the cells occupying that space in the immature plant die in accordance with a genetic program. The tail of a tadpole is reabsorbed through apoptosis.

There are indications that some human diseases such as Alzheimer's syndrome are caused by malfunctions in a programmed cell death mechanism[25].

The tiny (1 mm long) roundworm (*C. elegans*) is a favorite of genetics researchers partly because, unlike larger animals, every single cell is genetically programmed. Each adult worm has exactly 816 cells (excluding a variable number of gonadal cells) but during development exactly 131 cells are programmed to die. Researchers have been able to trace the entire cell division scenario from fertilized egg to each adult cell. *C. elegans* has a genome of 97 million bases, 6 chromosomes and 20,000 genes. The *C. elegans* genome has been completely sequenced.

In 2002 the Nobel Prize for Physiology or Medicine was awarded to Brenner, Solston, and Horwitz for identifying the genes associated with cell death in the roundworm as stated in their citation:

> "This year's Nobel Laureates in Physiology or Medicine have made seminal discoveries concerning the genetic regulation of organ development and programmed cell death. By establishing and using the nematode *Caenorhabditis elegans* as an experimental model system, possibilities were opened to follow cell division and differentiation from the fertilized egg to the adult. The Laureates have identified key genes regulating organ development and programmed cell death and have shown that corresponding genes exist in higher species, including man. The discoveries are important for medical research and have shed new light on the pathogenesis of many diseases."

If it were not for Darwin's dilemma, genetically programmed aging and other lifespan control mechanisms would be an obvious logical extension of programmed cell death. In fact, some researchers are exploring this avenue.

7. Modern Programmed Aging Theories

All of the scientific theories and popular opinions about aging fall into two categories, namely:

1) Aging is an evolved mechanism, an adaptation, of organisms similar to the ones that determine puberty age, provide for metabolism, accomplish sexual reproduction, or result in any species-specific physical property such as fangs or fur. Animals and humans are *designed* by nature and purposely genetically programmed to age. Aging provides some benefit to the organism and therefore has been selected. Aging is not a defect; it is a *feature* that has a purpose (e.g. Weismann's theory). Aging is the active result of a lifespan regulation mechanism.

2) All other explanations including the modern non-programmed theories. Aging is a *defect*, a fundamental property of life, or unavoidable adverse side effect of a necessary process. Aging is the passive result of forces acting upon the organism.

The choice presented by Darwin's dilemma is essentially between these two cases.

The main difficulty with the traditional non-programmed theories and other theories in category 2 is that the theories are too simple to explain the wealth of detail observed. None of the theories seems to fit all or even most of the observations. There are many theories. They cannot all be correct.

If aging is an evolved trait as defined for category 1, then fit is not as much of an issue. We are all familiar with the huge variety and complexity of evolved traits and with the fact that many of them seem to be bizarre in that the benefit of the trait is not obvious. Obviously evolved characteristics such as the physical designs of organisms are assumed to benefit the animal. Inherited behavior patterns (instincts) are assumed to (somehow) benefit the animal.

For example, nobody can prove that its tail benefits a rat. The tail can develop wounds and diseases. A tail requires resources. It must be fed. It adds to the weight that the rat must carry around and therefore the size of other structures and muscles. Can anyone actually prove that the benefits of a tail actually outweigh the costs? No. Alternately, the rats may have inherited their tails from some ancestor species. A tail may not benefit the rats at all but may have evolved because it benefited some ancestors in the same way that the human appendix is assumed to have benefited some ancestor. Tails on rats may be actually devolving.

Some animals are assumed to have traits that are not optimum for the animal but are assumed to be in the process of evolving. Some are assumed to have traits that are "throwbacks."

Bamboo, salmon, progeria, non-aging animals, the inter-species variations, and elephant teeth, all fit the "adaptive evolved mechanism" theory to the same extent that tails and other observed structure (all assumed to be evolved, adaptive traits) fit.

The main difficulty with the evolved mechanism theory remains the same as it was following the first proposals of such ideas by Darwin or Weisman: it conflicts with Darwin's theory of natural selection because it requires evolution of a trait that is adverse to individual fitness.

A second difficulty is the perception that aging in a wild population has a negligible fitness impact and therefore could exist despite the force of natural selection. Both of these issues will be discussed at length below.

A small but growing number of biologists and other theorists (including the author) believe that Weismann was right. Aging is an evolved adaptive characteristic that, while adverse to individual organisms, still provides an evolutionary benefit. These "adaptive" theories are all based on the idea that the orthodox theory of natural selection, although correct, is not complete and that therefore exceptions, additions, or adjustments are possible. Although Darwinian natural selection explains a great many things, the adjustments are needed to explain some things, including aging.

Aging is not the only characteristic of organisms that did not fit with Darwin's theory. As will be described in this chapter, theorists have been working for decades on proposing modifications to classical "orthodox" Darwinism in an attempt to explain some of the other discrepancies especially in the area of behaviors. Because aging was plausibly not an evolved characteristic, and because semi-plausible traditional (non-adaptive) theories existed to explain aging, this effort has concentrated on behavioral traits. Since behaviors are highly structured, it is less plausible that they could result from random processes.

Modern genetics has also disclosed aspects that appear to be incompatible with orthodox Darwinian theory as described under Genetics, specifically the "digital discrepancies."

Completeness of Orthodox Natural Selection Theory

Recall that in 1859 some people rejected the theory of natural selection because it was incompatible with observed lifespan characteristics. Today, many biologists reject adaptive theories of aging solely because they are incompatible with orthodox natural selection. Therefore, if an adaptive theory is correct then natural selection must be incorrect; if natural selection is correct, adaptive theories must be incorrect. Natural selection has endured for 150 years. Genetics has provided some independent confirmation of some aspects of natural selection theory. Few doubt natural selection. Therefore, QED, adaptive theories are incorrect.

Such black and white positions tend to ignore the obvious third possibility, namely that natural selection is *generally* correct but has an *exception* with regard to aging and some other organism characteristics.

Some people, including scientists, think that traditional evolutionary mechanics theory is at least as solid and certain as, for example, Einstein's theory of special relativity. There is actually almost no similarity:

-Relativity deals with relatively simple phenomena such as the motion of particles in a vacuum. Physics is "hard" science. Natural selection deals with extremely complex issues such as why all the living species exist and behave as they do. Biology is "soft" science. Evolutionary mechanics theory is "soft" even by comparison to other areas of biology.

-Relativity has been experimentally confirmed by thousands of experiments performed by hundreds of investigators. Much of natural selection theory cannot be explored experimentally.

-No one has discovered a single, repeatable exception to the relativity theory. There are many known and increasing discrepancies with orthodox evolutionary mechanics theory.

-By comparison, relativity is essentially a "fact." Orthodox evolutionary mechanics theory is only a "theory."

Despite this, some of today's traditional biologists feel that the theory of natural selection is not only correct but also so complete, comprehensive, and all-encompassing that any valid exceptions, additions, or extensions are impossible. For example, noted biologists Olshansky, Hayflick, and Carnes say in their *Scientific American* article *No Truth to the Fountain of Youth* published in 2004[26]:

> "The way evolution works makes it impossible for us to possess genes that are specifically designed to cause physiological decline with age or to control how long we live."

This statement was made after the discovery of aging genes in multiple organisms and publication of four different alternative evolutionary mechanics theories that provide theoretical support for "genes that are specifically designed to control how long we live." Note again the use of the word "impossible" (used previously by Williams in similar context), which is relatively seldom used in scientific papers. This paper is said to have been endorsed by 51 prominent scientists. These people think that it is impossible that their understanding of the evolution process is anything less than perfectly comprehensive. They think it is impossible that any future developments will alter that understanding. None of the proposed alternatives to orthodox theory consider that it is incorrect but only that it is incomplete and needs relatively minor modifications.

To assess the plausibility of their interpretation, we could consider the physics of motion, which high school physics students learn concerns inertia, action and reaction, $f = ma$, and so forth. Most biologists would (correctly) consider Newtonian mechanics to be essentially trivially simple when compared to almost any aspect of biology. Newtonian mechanics successfully explained at least 99.9 percent of the observations. Nevertheless, eventually, it was found that there was an exception. Particles approaching the speed of light did not follow Newtonian mechanics. Newtonian mechanics is correct. It works most of the time. It just is not complete. There is at least one exception.

Similarly, when it was determined that atoms were composed of the "fundamental" particles electrons, neutrons, and protons, a great many things were explained and it was tempting to say that we finally understood the composition of matter completely. Subsequently, we discovered that matter also included positrons, neutrinos, mesons, and a host of other "fundamental" particles. Again, it is still correct that matter contains electrons, protons, and neutrons; it's just not the whole story. Today very few physicists think that we totally understand the composition of matter.

In astronomy we have gone from the idea that "the Earth is the center of the universe" to the currently widely accepted idea that the universe is about 20 billion light years in diameter but was once (very momentarily) smaller than a golf ball!

Similar humbling examples could be cited from virtually any field of science.

In contrast, evolution is a process that has produced the most complex objects known to man, namely a huge variety of living organisms including humans. To consider that such a process could be completely, totally, henceforth, and for all time, described by a 150-year-old theory that could be written on the back of an envelope would appear to require an astounding level of scientific hubris. This sort of attitude is the scientific equivalent of a flea climbing an elephant's hind leg with rape in mind! The theory of natural selection is generally correct but is almost certainly not complete to the point that no exceptions, modifications, or additions are possible. Some of the exceptions, modifications, or additions that have been proposed and their impacts on aging theory are described in the following sections. All of these theories consider that natural selection is generally correct in that species evolved from other species, and in that individual survival and reproduction are the most important factors in determining whether a characteristic is evolved. However, they differ from orthodox Darwinism in that they all consider that *other natural factors* could also be involved in influencing the evolution of a particular characteristic.

The sort of attitude described above, that everything about the evolution process is already known, was common until the nineteenth century. However, because of events such as noted above, modern scientists in other fields almost never claim that their knowledge is totally comprehensive. The existence of non-science factors uniquely allows and encourages such views in evolutionary mechanics science.

Evolutionary Effects of Aging

One of the main objections to adaptive theories of aging is the relatively insignificant effect aging has on fitness because of the alleged "declining fitness effect of adverse events with age" (Medawar's hypothesis). Some traditional biologists dismiss adaptive theories for this reason and some biology textbooks heavily promote this idea. However, the traditional view of the effect of aging on fitness is overly simplified. In the following sections we shall see how characteristics of actual animals greatly increase the impact of aging on animal populations.

In the wild, animals die mainly of predator attack, warfare, inability to obtain food, disease, and environmental conditions.

Aging causes weakness, reduced agility and mobility, increased susceptibility to disease, increased susceptibility to adverse environmental conditions, deterioration of senses, and reduced reproductive effectiveness.

It is therefore clear that, in wild populations, well prior to the occurrence of "programmed death", aging causes greatly increased probability of death from the causes listed above. Aging does not have to, directly, by itself, cause death in order to result in death and thus have an impact on evolution. Aging just has to increase the probability of death from the listed causes. If a lion is chasing 100 wildebeest in the Serengeti which one is going to be caught? The one that is just a little bit slower or a little bit less lucky. After thousands of years, the luck part averages out.

Therefore, we can think of genetically programmed aging as causing programmed weakness, programmed increased susceptibility to disease, programmed reduced mobility, programmed reduced sexual vigor, etc. as opposed to "programmed death."

Because natural selection, acting during millions of years, can select between very small advantages or disadvantages, even a very small weakness, agility loss, or other deterioration, such as might occur in even a relatively young animal could have a significant effect on an animal's survival or breeding probability and thereby evolution.

The effect of aging in actual wild animals on fitness and thereby natural selection is therefore *not* insignificant.

Studies on large wild animals such as those of Anne Loison[27] of the Norwegian Institute for Nature Research (See Resources) confirm that death rates increase with age beginning at rather young ages.

Another aspect of the performance deterioration caused by aging in most animals is that it generally *gradually increases*. This is significant to some of the theories described below.

Species Semantics

Species in sexually reproducing organisms is generally defined for evolutionary purposes along the lines of "that group of organisms that can interbreed and produce fertile descendants." Members of a species can interbreed with each other even though they might be of different races or breeds. Members of different species cannot interbreed to any significant extent. Those familiar with different breeds of dogs know how different members of the same species can be from each other as a result of selective breeding. Conversely, different wild species can be quite similar physically. We now know that evolution of two different species from a single parent species (*speciation*) occurs when genetic differences between two breeds significantly interfere with the various complex processes involved in meiosis (such as matching and swapping of genetic instructions between parent chromosomes) and other aspects of sexual reproduction. Once speciation occurs, it is reasonable to believe that evolution of the descendent species proceeds more rapidly because interbreeding becomes difficult or impossible. Species is a very important concept to understanding evolution and to understanding the interaction of various organisms present at any particular point in time.

However, how does the concept of "species" relate to the time-sequential flow of evolution? It is wholly irrelevant if "mouse" of today could or could not interbreed with "mouse" of 1000 years ago. Assuming mice are under evolutionary pressure, then "mouse" of this year is presumably a minutely different species from "mouse" of last year. At what point is an evolving mouse a different species? Is "species" even an applicable concept when considering the time-sequential flow of evolution? Is "evolution of the species" an oxymoron?

Group Selection and Evolutionary Immediacy

All of the evolution theories we will be discussing agree that the design of an organism (or any complex system) is a forest of compromises or tradeoffs. Whether a design change would be incorporated by the evolution process depends on the existence of net benefit. A major functional difference in evolution theories is the extent to which a future, long-term benefit can offset an immediate, short-term cost during the evolution process.

As indicated earlier, one of the things that did not fit Darwin's theory was the evolution of some forms of bees, ants, and other colony insects. Since workers and warriors were sterile, they could not have evolved by means of strict Darwinian natural selection. Apparently, the workers and warriors evolved by virtue of the collective fates of their colonies. Depending on the beneficial characteristics of the workers and warriors, the *colony* would survive or not survive thus selecting the beneficial characteristics of the workers or warriors. This led to the idea that characteristics beneficial to a *group* could be selected.

An extension of this idea was to consider whether characteristics that favored survival of other groups could evolve. Could a characteristic that favored a herd or other animal colony evolve?

Next, could a characteristic evolve which was favorable to a group even though it was *adverse* to individual animals? Could a tradeoff exist between group benefit and individual disadvantage?

Finally, could a characteristic evolve which was adverse to individual animals but favored the species, i.e. *species-level group selection.*

The later cases are different from the case with ants and bees because the individual animals can reproduce. Therefore, group survival is competing with individual survival. People who believe in group selection believe that if the group benefit is sufficiently large, and the individual survival or breeding disadvantage is sufficiently small, a characteristic benefiting a group can be an evolved characteristic.

There are animal characteristics besides aging that do not appear to obey the rules of Darwinian natural selection and fitness, especially in the area of behaviors. For example, *altruism* is a tendency of an animal to behave in a manner not consistent with its own best interest from an orthodox fitness point of view.

Some animals will protect the young of other, unrelated, animals of the same species. It would make more sense from a fitness point of view to attack or at least ignore the young of an unrelated animal. Protecting unrelated young puts an animal at risk thus reducing its chance of propagating in favor of increasing the survival chance of an unrelated, nominally competing, animal.

However, protecting unrelated young might well provide a group benefit for the herd or species by increasing the collective chance of survival.

Another behavioral characteristic that is troubling to Darwinian theory is the relative lack of aggression between members of the same species. According to "dog-eat-dog", "red of tooth and claw" orthodox Darwinian theory, competition is fiercest among members of the same species. Members of the same species have, by definition, identical requirements for food supply and habitat and therefore should be in stronger competition with each other than with members of other species whose needs do not overlap as much. Yet many instances of fighting between members of the same species seem ritualistic and designed to determine pecking order, territorial rights or mating rights rather than inflict permanent harm.

There are a number of similar examples in the behavior area.

This sort of analysis introduced a question that has now been bothering scores of theorists for at least fifty years:

> *Regarding the evolution process, what is the relationship between the survival/ reproduction or non-survival/ non-reproduction of individual members of a species population and the extinction or non-extinction of the population?*

It is obvious that an extinction event changes the subsequent biosphere. The extinct population vacates habitat and frees resources that can be used by other populations. The extinct population does not subsequently produce any descendants or descendant species. If the immediate ancestor species that produced the human species had previously become extinct, humans would not exist. If the dinosaurs had *not* become extinct, perhaps humans would not exist now. The question concerns the relative importance of population extinction and individual survival *in directing the evolution process*. Let's break this question down into smaller questions:

> *Can we imagine a class of organism traits that increase the probability that a population would survive but are simultaneously at least slightly adverse to the survival and reproduction of individual members of that population?*

It is clear that in most cases traits that help individuals survive and reproduce would be good for a population of those individuals. However, human societies and civilizations are rife with laws and commandments that limit the actions of individuals in favor of the group and apparently similar traits have been observed in non-humans. "Thou shalt not covet thy neighbor's wife" might be good for a society but obviously not good for the reproductive potential of the constrained individuals. Theorists have suggested many organism characteristics that could benefit populations at the expense of individuals, including internally purposely limited lifespan as will be discussed in detail. There does not appear to be significant scientific opposition to these ideas, per se.

For example, imagine that individuals in a population of animals were so good at surviving and reproducing that they could cause an overpopulation, which in turn was decimated by starvation and disease. Their habitat and food supply could then be taken over by competing populations. Now imagine that these animals had a variant that could detect overpopulation and limit their individual survival and reproduction in order to avoid overpopulation and subsequent adverse impact to the population. Such a trait would be individually adverse because it would reduce the probability that any given individual would produce descendants but would plausibly benefit the survival of the population.

Can an organism trait that benefits the probability that a population will survive at the expense of individual survival/ reproduction be incorporated into an organism's design via the evolution process?

Of course this is the big question. Theorists have divided into essentially ideological factions on this issue. Keep in mind that the traditional aging theories based on Medawar's idea are at least somewhat population-oriented as described earlier. Huge ideological arguments persist between various levels of population-oriented vs. individual-oriented theories.

The term ***group selection*** is used to denote a theory that proposes that group benefit can override some amount of individual disadvantage and that a tradeoff can therefore exist between group benefit and individual disadvantage. Group selection was described by Wynne-Edwards in 1962[28]. Other theories such as the selfish gene theory and evolvability theory (to be discussed) also suggest that group or species benefit (or benefit to the evolution process) could override individual disadvantage but provide for somewhat different evolutionary mechanics.

A very central concept to evolutionary mechanics theory is that of *immediacy*. The nominal presumption is that a "near-term", "immediate" (i.e. individual) benefit would have more impact than a comparable "longer-term", "group" benefit, which in turn would have more impact than a very long-term "species" benefit. Any discussion of group selection invariably involves discussion of short-term disadvantage vs. long-term advantage. The extent to which group selection is feasible depends on one's conception of what we might call an "immediacy factor", the degree to which long-term is less significant than short-term.

Orthodox Darwinists think that this immediacy factor is so large that any sort of group selection is impossible. Group selection was severely criticized by G. Williams[29] (author of the 1957 orthodox-based antagonistic pleiotropy theory) and other traditionalists in the late 1960s. Some traditionalists consider that group selection has been "definitively" defeated by these arguments. Wynne-Edwards subsequently (1986) rebutted[30] and the argument continues much less vigorously to this day. Some group selectionists think group selection is possible but generally only in relatively short-term circumstances such as small groups. Presumably the larger the group the longer it would take for a group benefit to be felt. ***Kin selection*[31]**, in which the group is confined to closely related individuals is an example of small-group selection.

Group selection involves another layer of complexity. Instead of just worrying about tradeoffs between individual benefits and individual costs, we are now additionally concerned about tradeoffs between short-term costs and long-term benefits.

The plausibility of group selection depends on the assumed time-frame or immediacy of the evolutionary process itself. By evolutionary process is meant the process that determines the evolved, accumulative, design of an organism. If evolution is a very long-term process, it is plausible that there would not be much, if any, difference between short-term and long-term benefit, so long as the "long-term" was shorter than the time required to determine "benefit." If evolution is a short-term process, such that new beneficial characteristics can be quickly incorporated into an organism's genome relative to speciation, we would expect a large difference.

People who think of evolution in more philosophical terms tend to be relatively unaware of this issue. A design feature that provides net benefit can evolve, regardless of time-frame. People who think in terms of the actual sequence of events attributed to Darwinian evolution (beneficial mutations, propagation, radiation, etc.) tend to believe in a (relatively) short-term evolutionary process.

Thousands of years of selective breeding certainly suggest that organism design change can be rather immediate. We can nominally selectively breed organisms to emphasize or attenuate any inheritable characteristic that varies between individuals of a species. If we selectively breed an organism we can indeed cause large changes in its design within a few generations. The mechanics of Darwinian evolution (based substantially on selective breeding) reinforce the idea that the immediacy factor is large and that thus any group benefit would have to be enormously greater than the individual disadvantage, *and* the group relatively small for group selection to work.

There is also little doubt that we could improve the fitness of a simple organism such as a bacterium by selective breeding. If we selectively bred bacteria for resistance to a particular antibiotic, their resistance presumably would improve. If we released these altered bacteria into the wild they would be highly competitive in a world that included the antibiotic. We could reasonably expect that the altered bacteria would become dominant in the wild population.

However, in more complex organisms there are some counter arguments. Remember that survival and fitness depend on the *combined net effect* of *all* of an organism's expressed design characteristics. We *also* know from selective breeding experience that selectively breeding complex organisms for any *one* characteristic invariably inadvertently introduces changes to *other* characteristics, all of which are nominally adverse. Nature can eventually sort this out by adjusting other characteristics to produce a net fitness benefit. The sorting process is clearly a much longer-term process in more complex organisms.

Second, we now know that in complex organisms at any given time, only a tiny portion of the inheritable genetic data (~0.3 percent in humans) varies between individuals and therefore selection at any one time is confined to the small fraction of characteristics controlled by that tiny portion. This is a severe limitation on the evolution process because the "mix and match" of different combinations of the variable design elements is limited to the small portion. The slow introduction of new propagatable mutations contributes to the long-term process by producing changes to the vast majority (~99.7 percent in humans) of genetic data and corresponding characteristics that can then participate in the sorting process.

If we took some antelope out of the Serengeti and selectively bred them for speed we could produce a faster antelope. If we release these new antelope in the wild would they successfully compete with the old antelope? Almost certainly not. The inadvertent adverse characteristics would outweigh the speed increase. The new antelope are "domesticated" in that they are bred for a specific quality at the expense of other qualities. Evolution is even slower than it appears. *Change* is not the same as *evolution*.

In the immediacy spectrum, orthodox Darwinism is at the short-term end, kin selection, small-group selection and the selfish gene theory are medium-term, and species-level group selection theories are long-term.

In this book you will see the rather awkward locution "non-individual benefit" or "diffuse benefit" as opposed to "individual benefit" and encompassing group, small group, evolvability, and population-oriented theories such as the selfish gene theory. Is not a non-individual benefit obviously a group benefit? The reason for the awkward phrasing is that there exist a number of slightly different group theories whose authors would no doubt object to being summarily lumped together. In addition, the mechanics of evolvability are substantially different from those of group selection as will be described.

Joshua Mitteldorf is an evolutionary biology theorist affiliated with the University of Arizona. Like the author, Mitteldorf has extensive experience outside the field of biology. He strongly believes that aging is an evolved characteristic by virtue of group selection.

Mitteldorf has published many papers[32] attacking the traditional theories of aging based on discrepancies between observations and the predictions of the theories.

Mitteldorf thinks[33] the benefit of aging results from population dynamics:

> "Ageing has evolved based on its contribution to stabilizing population dynamics, helping prevent population growth overshoot, exhaustion of ecological resources, and local extinction. … Ageing would become one of the mechanisms by which a species can take control of its death rate, suppressing violent fluctuations that might otherwise cause extinctions. Ageing (together with reproductive and predatory restraint) would help hold the growth rate down below the chaotic threshold."

These benefits of aging are discussed at length in the following sections.

Group selection as a rationale for an adaptive, evolved, aging mechanism generally has difficulties regarding *mechanics* as described above.

In a paper[34] published in 2004 titled *The Evolution of Programmed Death in a Spatially Structured Population*, Justin Travis presented a mathematical model in which programmed death as an adaptation could evolve. This concept is similar to the "small group" model. However, Travis' model requires that a non-aging species have a reproductive capacity that declines with calendar age. As indicated earlier, such a requirement has its own theory problems.

J. Bowles is another theorist with experience outside traditional biology. Bowles published a paper[35] in 1999 called *Shattered: Medawar's Test Tubes and their Enduring Legacy of Chaos* that extensively criticizes logical flaws in the development of the mutation accumulation theory including some of the problems with the traditional model of non-aging animals described in this book. Bowles also believes that aging is an evolved trait by virtue of group selection and has theorized extensively on the relationships between aging and sex, aging and availability of food, aging and presence of predators, and other indications that aging is part of a larger, more comprehensive, deliberate mechanism with a purpose.

Some of the relatively recent discoveries in genetics (see following section) support group selection by suggesting that evolution of the genetic code is a very complex long-term process (even by evolutionary standards) and that evolutionary modification of some parts of the code (and their resulting characteristics) could take much longer than other sorts of modifications.

Traditional biologists take the position that because we do not completely understand it, it cannot exist, despite all the evidence. Group selectionists take the position that eventually we will understand it.

Aging has semi-plausible (traditional) alternate explanations. Psychology is "soft" science. Behaviors are subject to interpretation. Some dismiss group selection as a misguided effort to ascribe human societal behaviors to animals.

Complex Evolutionary Processes

In chapter 5 we discussed the major differences that exist between the inheritance process of simple organisms and complex (sexually reproducing) organisms, differences that act to change the way evolution works in complex organisms and specifically produce a longer-term process. Here we discuss additional reasons for believing in long-term evolutionary processes in complex organisms.

Most people would summarize the mechanics of the evolutionary process in Darwin's theory approximately thus:

> Random changes to an organism's inheritable characteristics occur. Occasionally, such a mutation is beneficial to the organism's ability to survive or breed. The ***individual*** organisms ***possessing*** the mutation survive longer, breed more, and therefore that mutation becomes more prevalent in the gene pool. Genetic isolation eventually leads to speciation. Evolution is a slow process because it is very incremental and because beneficial mutations are very infrequent. However, once a beneficial mutation occurs, the beneficial effect is ***immediate*** in those organisms possessing the mutation.

Orthodox Darwinian evolution depends on the idea that a single mutation can immediately provide increased survival benefit to an individual organism that *possesses the mutation*, and that this mechanism is the only force behind evolution. Each mutation is evaluated individually, on its own merits by the natural selection process.

This concept of *immediate single mutation benefit* (also known as the one-mutation-at-a-time scenario) is directly incompatible with group selection and all the longer-term theories. If evolution takes place by means of mutations that cause individual organisms to live longer and breed more, then obviously a mutation that causes lifespan to be reduced (without some compensating individual benefit) or otherwise results in a net individual disadvantage presumably, essentially by definition, cannot result in an evolved characteristic, regardless of any group benefit. Immediate single mutation benefit leads to the concept of "individual" fitness. Propagation of the beneficial mutations requires that the organisms possessing them live longer and breed more.

In chapter 1, we saw where Darwin concluded that only mutations that caused small changes (nearly neutral fitness effect) were likely to be beneficial. Because "benefit" is the result of the combined effect of all of an organism's characteristics, changes in characteristics need to be highly coordinated to produce a beneficial effect. This could only occur if the changes were incremental.

Changing any characteristic of an organism therefore presumably interacts, to some varying extent, with *all* of the organism's other characteristics. As organisms become more complex, the number of potential interactions increases with the factorial of the number of characteristics and therefore the probability of a single random change being beneficial decreases dramatically with increasing complexity.

More advanced organisms are also more tightly *integrated*. That is, their various parts have an increasingly greater dependence on each other and increasingly critical relationships with each other. For example, a jellyfish, earthworm, or plant might well survive a major physical injury that would be immediately fatal to a mammal. As complexity *increases*, the magnitude of a potentially beneficial change would therefore have to *decrease*.

There are other reasons, described in the following sections, that evolution becomes progressively more difficult, and therefore should be progressively slower, as complexity increases. In actuality, evolution appears to have become more rapid as organisms have become more complex.

Orthodox mechanics would therefore appear to have limitations regarding complexity. Is there a maximum complexity, beyond which evolution cannot proceed? If evolution depends on beneficial changes that become increasingly minor as complexity increases, would there not come a point at which changes causing disadvantage would overwhelm those causing advantage?

Our understanding of the digital nature of genetic data suggests another problem. Digital data is not continuously variable. Because of granularity, there exists a finite minimum size "step" to the smallest possible change to any given parameter that could be caused by a single mutation. Would not this "quantum effect" cause a limit to the complexity that could be supported by Darwinian evolution? The quantum effect prevents individual changes from becoming indefinitely more minor as complexity increases.

Chapter 5 revealed another problem with the single mutation immediate benefit concept in more complex organisms: The magnitude of variations in survival traits due to mutation is much less than that due to recombination of existing mutations. Therefore, it would appear that recombined mutations were more important to the evolution process than the underlying individual mutations. This is a very important point. Although mutational changes are the raw material of evolution the natural selection process in complex organisms actually operates on selectable properties of organisms that result from combining mutational differences. It is very important to understand the differences between *mutational differences* and *selectable properties*:

Mutational difference: An inheritable genetic change (mutation) that occurred in a single individual and then propagated to a substantial portion of the population; typically a single nucleotide (one-letter) difference in the organism's genetic code. In the total population of a complex species (such as a mammal) there simultaneously exist millions of mutational differences. Any two, even closely related members of a complex species, are likely to have thousands of mutational differences between them. In order to have propagated, a mutational difference would necessarily have to result in only minor (positive or negative) phenotypic effect. There can be thousands of differences between the two genomes possessed by a *single* individual!

Selectable Property: A design aspect of an organism that could plausibly be selected as "beneficial" by natural selection (e.g. speed, strength, height, visual acuity, intelligence, ability to climb trees or elude predators or catch prey, etc. etc.). Because of recombination and the diploid nature of sexually reproducing organisms, descendants of a single pair of individuals, all resulting from the same four sets of genetic data, can have very large differences in their selectable properties. This is a major contributor to local variation.

In a simple organism (bacterium) a single mutational change could plausibly result in a selectable property. In a complex organism, this would almost never be the case. In a complex organism, selectable properties result from particular ***combinations*** of mutational differences. Mutational difference ***is not*** the same as a selectable property.

Here is a thought exercise illustrating this issue:

Imagine that in some animal population a particular height was optimum. Of the millions of mutational differences (single nucleotide polymorphisms) that existed in this population imagine that 100 affected height. We could label these H1 through H100. For each of these polymorphisms we could designate the allele that results in a taller individual as t and the other allele corresponding to the shorter configuration of that polymorphism as s. We then have H1t and H1s, H2t and H2s, etc. Suppose we measure the heights of many animals and find that their heights fit the usual bell-shaped curve. The average animal probably possesses approximately half of its alleles in the s configuration and half in the t configuration. Taller animals have more t alleles, shorter ones have more s alleles. Now suppose we selectively breed this animal for "tall." We could eventually create animals that are taller than any in the original population because they have more t alleles. However, there would be an ultimate limit in height represented by a case in which an animal possessed all of its 100 H polymorphisms (in both of its genomes) in the t configuration. Eventually, new mutational changes would occur that would add to the number of height polymorphisms.

If conditions changed in the wild population of this animal favoring taller individuals, natural selection would operate in the same manner. Mutations are the feedstock to the evolution process but natural selection actually operates upon selectable properties. This has implications for evolution theory.

More generally, in Chapter 5 you read that, subsequent to Darwin, the inheritance process in more complex organisms has been found to involve many complex mechanisms in addition to natural selection that clearly affect the overall process of evolution. These evolutionary mechanisms include sexual reproduction, more effective recombination of genetic data, unequal crossover, existence of and mechanics of diploid chromosomes such as pairing and meiosis, and evolutionary aspects of the organization of digital data in genomes.

The process of evolution in complex (diploid) organisms is therefore very different from that of simpler organisms. Consider a mutation to genetic data that causes a minor adverse fitness effect. Because diploid organisms contain two sets of genetic data, the adverse effect might not be fully expressed in an organism in which the second set of genetic information did not contain the mutant data. If the mutant trait was recessive, an organism possessing a non-mutant set as well as a mutant set of data would not express any adverse effect. Diploid organisms can *possess* the adverse mutation without *expressing* the adverse trait. In a haploid organism, any organism that possesses an adverse mutation also expresses that adverse trait. Therefore, mildly adverse mutations can propagate further and be retained in a species population longer in a diploid organism than in a haploid organism. Likewise, a beneficial mutation would propagate less rapidly in a diploid organism. This feature of diploid organisms supports maintaining genetic diversity and also supports adjustments to Darwinian evolution as described below. It is also one of the observations that are directly incompatible with traditional evolutionary mechanics: Why would sexual reproduction evolve and be retained if it is adverse to the propagation of beneficial mutations and encourages propagation of adverse mutations?

Although random changes in an organism's digital genetic data are no doubt the "input" to the evolutionary process, a number of evolutionary mechanisms in addition to natural selection process, filter, sort, and organize the random data changes as part of the overall process of evolution in more complex organisms.

While evolution in very simple organisms such as bacteria could plausibly proceed in a more or less Darwinian manner, evolution in complex organisms is a much longer and more complicated process than could possibly have been anticipated by Darwin.

The evolutionary mechanisms are obviously evolved. They are complex, highly structured, and do not exist in the simpler organisms.

These mechanisms also themselves appear to be generally incompatible with orthodox Darwinism. For example, recombination is individually adverse. An animal possessing a beneficial mutation cannot depend on that mutation being passed to its descendants. These issues are discussed further in the following sections.

Many of the mechanisms that cause plausible evolutionary impact have no individual fitness effect. These include the organizational changes in genomic data that we could include under the term "evolution of the genome" including transposition of data within a chromosome, insertion of introns, changes to junk DNA, and so forth. The genetic communications system itself has a complex design. The design of this system is independent from fitness just as the mechanism and language of an email do not affect the content of the message.

Many of these mechanisms operate over time scales that are very large compared to the time scale under which speciation occurs, much less the time scale at which individual mutations occur. In orthodox Darwinian evolution, we could imagine that a mutation to data in a sperm cell would *cause* a beneficial change in the very animal produced by that sperm cell. In complex evolution, because of all the recombining, transposing, and clustering, we could imagine that a mutation might not *participate* in a beneficial effect for millions of years after it occurred.

In Darwin's "analog" world, it was a reasonable assumption that all characteristics of organisms were continuously variable and therefore equally subject to the force of natural selection. Our current knowledge of the structure of the (digital) genetic code indicates that certain characteristics could be essentially immune to natural selection relative to other characteristics and that some parts of the code and their resulting characteristics could (and do) have lifetimes much greater than a species lifetime. Apparently, at any particular point in time only a tiny portion of a genome varies between individuals and therefore only that tiny portion is subject to natural selection. A species could therefore inherit a "species benefiting" characteristic from an ancestor species that was robustly resistant to out selection and would therefore survive the tendency to select out because of individual disadvantage.

For all of these reasons, it is reasonable to believe that immediate single mutation benefit does not represent the primary mode of evolution in more complex organisms. Instead, mutations that are individually fitness-neutral or mildly adverse occur and are distributed rather widely in a population. Individual organisms, created by recombination, and possessing beneficial *combinations* of these mutations live longer and breed more. The nearly neutral mutational alleles underlying the beneficial combinations are therefore propagated more widely. Other, longer-term mechanisms cause genome organizational changes that affect subsequent evolution. Individuals do not "possess" beneficial mutations. Evolution is not driven by beneficial mutation. Mutations are only beneficial in combination with other mutations. Mutations do not have immediate beneficial effect.

One clue that supports this concept is the nature of variation. The extent of variation (in fitness terms) that exists in a population due to recombination is usually much larger than that resulting from any individual underlying and plausibly beneficial mutation.

If immediate single mutation benefit is not the primary force behind evolution in more complex organisms, then the main conceptual barrier to longer-term evolution theories is removed. There is no "individual fitness", just "fitness." A tradeoff between a long-term benefit and an individual disadvantage would appear to be as feasible as any other tradeoff. The existence of complex evolutionary processes therefore supports long-term theories for more complex organisms.

The existence of evolved mechanisms that support and improve the process of evolution leads directly to *evolvability theory* (to be described). If organisms can evolve *some* characteristics that help them to evolve, what *other* such characteristics might exist?

Where group selection has tended to be based on behaviors, and therefore subject to endless argument, emerging genetic mechanisms are relatively "hard" science, confirmed by repeatable experiments performed by independent investigators. We can therefore hope that further developments in genetics will eventually definitively settle questions about evolution, evolvability, and group selection.

Evolution (and evolution theory) is all about propagation. We could imagine that at some point in time (let's say point "A") a mutational change occurs in a single member of a sexually reproducing species. Now suppose that at some later time, point "B" this mutational change has propagated to essentially the entire population and is now part of the normal genome for that organism. We can all agree that natural selection *differentially* affects mutational changes on their journey from point A to point B and that therefore natural selection processes and filters the mutational changes. We can also agree that these mutational changes must pass through the organism's inheritance system multitudinous times on this journey. Therefore if the inheritance system has any features or properties that *differentially* affect mutational changes that these features and properties would ***also*** act to filter and process mutations. This book describes how many inheritance features (paired chromosomes, genetic linkage, pattern sensitivity, unequal crossover, transposons, etc.) do differentially affect mutations and how these features interact in very complex ways. Example: If a mildly individually adverse mutational change is located on the same chromosome and physically near a group of other mutational changes that produce a beneficial effect, then propagation of the first change will be enhanced because of genetic linkage.

If someone tells you that they understand evolutionary mechanics or that they have a model for the evolution process, ask to see how their understanding or model deals with each of these features and further how they deal with interactions between differentiating inheritance features and natural selection. You will likely be disappointed by the answer. Be especially wary of those making derivative pronouncements (e.g. "Group selection is impossible because *my* propagation model says it's impossible.").

Inheritance Efficiency and Individual Advantage

Darwin's mechanics concept is based on the idea that organisms possessing traits that improve their ability to survive or breed *pass those "survival" traits to their descendants.* We recognize that a trait that was not genetically recorded cannot participate in the evolution process. In fact, "acquired" traits that are *not* genetically transmitted but *are* important to survival (such as knowledge and experience) can have a negative effect on evolution as explained in following sections.

It follows that the *degree* to which an individual can transmit its traits to descendants (we could use the term *inheritance efficiency*) is important. In other words, to what extent do its descendants resemble their parent? This aspect of individual advantage is not contained in the concept of fitness.

As an illustration, we could consider a "limit" case. Suppose some mutational change allowed a duck to produce twice as many eggs as before, clearly a large fitness advantage. Now suppose that this change somehow also caused a situation in which none of the parent's characteristics were transmitted to these descendants. All of this duck's eggs somehow produced geese. Its descendants do not express any of its personal characteristics. The eggs hatch just as well as they did previously. Mortality in the young birds is no worse than before. Fitness is improved. However, this mutational change represents a catastrophic individual disadvantage. Even though "fitness" was improved, this duck cannot pass its characteristics to its descendants. This concept of *inheritance efficiency*, the degree to which an individual's characteristics are transmitted to descendants is important to adjustments to Darwin's theory such as the selfish gene theory and evolvability theory.

In this connection, we know that a diploid sexually reproducing organism passes nominally 50 percent of its genetic data to each of its descendants. Does this fixed *amount* of data correspond with a fixed degree to which its descendants express its survival characteristics? Is inheritance efficiency a fixed factor in sexually reproducing organisms? Can we therefore ignore this factor because no sexually reproducing organism has more or less inheritance than another? Because of the mechanics of recombination, the answer is "no" as illustrated in the following sections. The important thing to evolution is the degree to which the parent's *survival characteristics* are passed to descendants, not the *amount* of genetic data transmitted.

Genetic Diversity and Individual Advantage

It is clearly in an individual's orthodox "Darwinian" interest to have the greatest possible inheritance efficiency in the transmission of survival characteristics to descendants. The more resemblance there is between an individual and its descendants, the more likely it is that beneficial traits possessed by the individual will be present in its descendants.

If an individual could choose its own, most advantageous, method of reproduction, it would opt for cloning. The inheritance efficiency of a clone is nominally 100 percent. All of the parent's inheritable characteristics would be preserved in its descendants.

Failing cloning, an individual would want to mate with another individual that was as nearly identical to itself as possible. Such a procedure would tend to minimize the extent of differences between itself and its descendants. Genetic diversity is therefore individually adverse.

Genetic Diversity and Evolution

However, as Darwin tells us, *evolution* is driven by *variation*. Natural selection depends on *differences* between individuals in a population.

In another limit case we could consider what would happen if the entire population of some species consisted of identical clones, genetic duplicates, of a single individual. Natural selection (or selective breeding) would not work in this population because there are no genetically transmittable differences between individuals for natural selection (or human breeders) to select. Although these individuals might be as fit as the members of some other population, and although their species as a whole might be competitive with other species, they would be unable to evolve further in the manner available to other, normal species that did possess individual variation. More genetic diversity causes more variation and therefore aids evolution. From an evolution standpoint, an individual should mate with another individual that is as different as possible from itself.

Therefore, there is a conflict and apparently a necessary compromise between the needs of the evolution process itself and individual advantage. Darwin could well imagine that variation was caused by some non-changing fundamental property of life. Modern "digital" genetics shows that variation is actually produced by evolved characteristics that are different in different organisms. The idea that the evolution process itself varies and can be affected by evolved characteristics leads directly to evolvability theory (to be described).

Notice that in the examples given above, subtle characteristics such as behaviors in choosing mates ("sexual selection") could favor either the evolution process or individual advantage in more complex animals. If an animal had an inherited behavior pattern that caused it to prefer mating with close relatives, that behavior would favor individual advantage.

If it had a behavior that led it to seek mates that were less related, that behavior would favor evolution at the expense of individual advantage. The second case is individually adverse for another reason: Presumably, close relatives are physically closer and therefore easier to find. An animal mating with a close relative is therefore less likely to die before finding a mate. Most of the more complex animals actually do have complicated mating behaviors that generally appear to avoid mating between close relatives.

The above discussions disclose weaknesses in the individual fitness concept that will be discussed further in sections to follow.

Genomic Design, Evolutionary Rigidity, Pleiotropy, and Group Selection

Everybody recognizes that the phenotypic future of an organism is largely dictated by its current design. Evolution by definition is incremental. The potential path followed by future evolution of an organism is largely constrained by its past. However, it is now increasingly recognized that the ***genome*** of an organism also has a particular design and that this design *also* incrementally evolves and additionally constrains the path of future evolution.

Darwin imagined that mutations happen and natural selection selects among the mutations, a simple and elegant idea formulated prior to any understanding of inheritance mechanisms. However, as shown in below and described in Chapter 5 intervening genetics discoveries have revealed that many different processes are involved in genome evolution and that these processes operate over dramatically different time scales.

Goldsmith

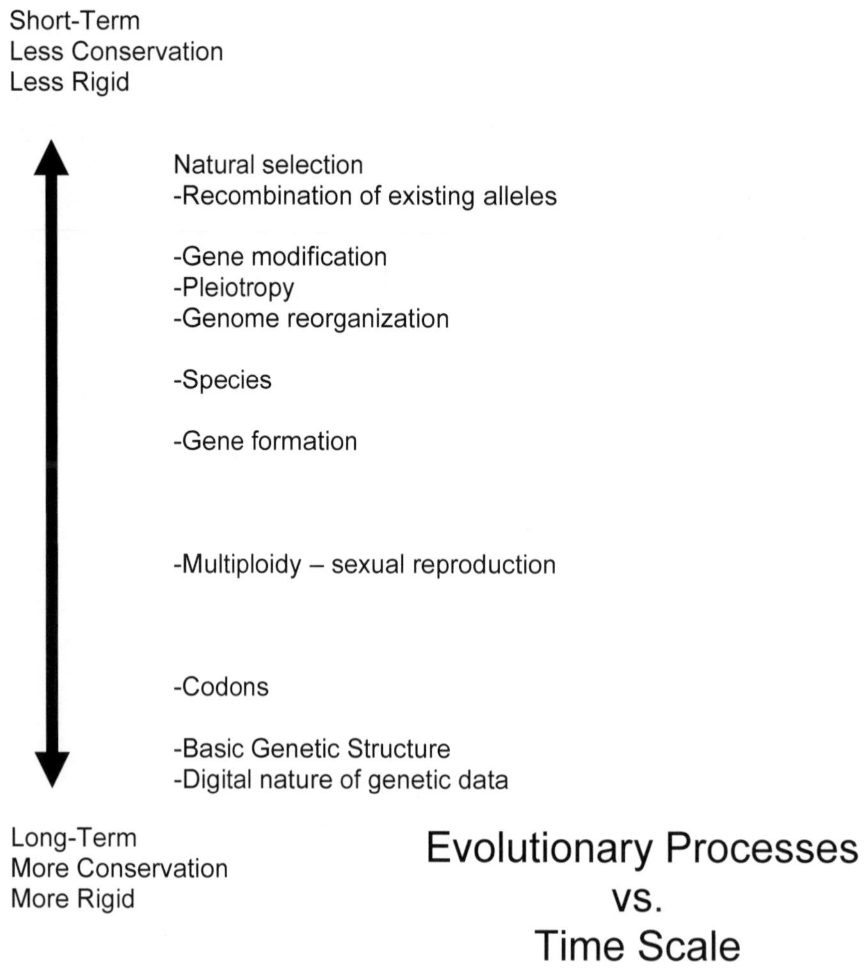

Figure 23 Evolutionary Processes vs. Time Scale

Imagine that some change in an animal's external world requires a particular organism design change. Suppose an anteater needs a longer snout and tongue because ants are building deeper nests. Nature's task is to modify the snout design *without changing any other phenotypic parameters* as these are presumably already nominally optimum. If this change can be accomplished by merely recombining alleles that already exist in the local population, we all agree that the change can be accomplished in a very short time (by evolutionary standards). This is the sort of change that could be accomplished by selective breeding and could occur in a few generations.

However, we now know that only a tiny fraction (~0.3 percent) of mammal genetic data varies between individuals and that therefore the variety of changes that can be accomplished by merely reassembling existing alleles is very limited. Even if snout length is one of the parameters that happen to be affected by the variable fraction of genetic data, the range of achievable phenotypic change might be very limited. Further, selectively breeding for longer snouts would likely introduce changes to other phenotypic parameters, all of which are nominally adverse. This limitation, *which would also apply to natural selection*, is an example of *linkage* between *phenotypic* design parameters caused by *genomic* design.

If the change required a specific new mutation to a particular gene, a very much longer time would likely be required. If the change required specific complementary changes to many different genes, a yet longer time scale would be involved. If the change could not be accomplished without the creation of an entirely new gene, a yet longer time regime would be invoked. For various theoretical reasons, the creation of a functionally different gene is an extremely difficult event, even relative to those already mentioned and yet it is obvious that, at some point in the evolution of complex organisms, new genes would be required. The conservation of genes and the time scale involved here is thought to be longer than typical species life (time since the species diverged). This is the basis of the gene-centered evolutionary mechanics theories such as the selfish gene theory described below.

Therefore, we can read down the rigidity chart in order of genomic design aspects that are increasingly more fundamental, therefore more conserved, and more rigid or increasingly difficult to change. Near the bottom are aspects that have been nearly completely conserved during the evolutionary life of the Earth (e.g. codon definitions). At the bottom are fundamental unalterable aspects (e.g. consequences of the digital nature of genetic data). The *rigidity* of linkages depends on the degree of difficulty involved in removing them. *Robustness* and *plasticity* are other terms used to describe the difficulty and therefore evolutionary time required to accomplish a specific change or resistance to specific changes.

Pleiotropy refers to the fact that a single gene typically controls multiple phenotypic properties. A single change to a single gene would typically alter more than one phenotypic property, thus displaying one specific type of genomic linkage. Conversely many phenotypic properties are directed by multiple genes. Williams (1957) suggested pleiotropic linkage between aging and some unspecified individually beneficial qualities as the reason why aging would not have been selected out despite its individually adverse nature.

One of the problems with this idea is that according to the underlying value-of-life concept, a longer lifespan has always been individually beneficial, at least since the development of sexually reproducing organisms. Nature would therefore have had a very long time (~3 billion years) to overcome the pleiotropic linkage. A pleiotropic linkage can not be indefinitely rigid. Complementary changes to many genes might well be able to accomplish the beneficial phenotypic function without the side-effects. If not, new genes could be created in order to overcome the linkage. Therefore pleiotropy would appear to operate over a time frame shorter than typical species life. Why wouldn't the pleiotropic linkage have been overcome? Such linkages did not prevent the anteater from obtaining a longer snout or prevent different mammals from evolving grossly different phenotypic designs. Pleiotropy represents a degree of rigidity that is short-term relative to a typical mammal species life time. Therefore pleiotropy does not appear to be a valid reason for explaining why an individually adverse design characteristic (aging) would not have selected out over a very much longer time.

The pleiotropy argument can be reversed as follows to favor group selection given the concept that group selection is relatively long-term and "slower" than individual selection. If a pleiotropic linkage between a group-benefiting but individually adverse design characteristic (e.g. aging) had formed in the primordial past, the linkage would tend to prevent aging from being selected out in the short-term, while the group benefit of aging would prevent it from being selected out in the long-term. In this concept, pleiotropy need not be perfectly rigid thus overcoming the difficulty mentioned above. According to this concept, antagonistic pleiotropy works better for aging as a group-benefiting but individually adverse design characteristic than it does for a characteristic (e.g. aging) that, according to traditional theories, has always been individually adverse and has never had any evolutionary benefit!

Pleiotropy is *only one* of the aspects of genomic design that plausibly impact evolutionary mechanics theory. The above discussion illustrates how relatively recent genetics discoveries have introduced major issues into thinking about evolutionary mechanics.

The Selfish Gene Theory of Evolution

Darwin's theory proposes survival and reproductive capacity as the factors driving evolution. However, Darwin makes it clear that the way these factors work is to increase the probability that an organism's inheritable characteristics will be propagated to a larger number of future organisms. We could therefore simplify and restate Darwin's theory to read that *any* characteristic can be evolved that aids an organism in propagating its genes.

Zoologist Richard Dawkins developed and popularized this approach in his 1976 book *The Selfish Gene[36]*. Dawkins wrote this book at least partly in order to debunk group selection theory by showing that some troublesome evolved traits such as altruism could be explained more or less within Darwin's theory ("Neo-Darwinism") where group selection represented a relatively more gross violation of Darwin's mechanics.

In Dawkins' view, it is the genes that are actually competing and struggling for survival. Where individual organisms only live for momentary periods, genes "live" for millions or even billions of years. Animals and humans are "survival machines" whose purpose is to propagate genes and whose every characteristic is evolved to propagate their underlying genes. Dawkins' view is essentially a rationale, a justification, for believing in a longer-term evolution process.

You will recall that altruism involves an animal protecting or otherwise helping an unrelated animal. Although an animal protecting its mate or progeny does not help the animal itself survive or reproduce further, most people would agree that such behavior helps the animal propagate its characteristics by increasing the probability that the animal's own progeny will survive. The tradeoff is between the increased survival risk to the parent (bad) and the decreased survival risk of the progeny (good). Such behavior is not really "altruism."

What happens if an animal protects progeny of another animal, maybe a niece or nephew as has been observed? Now it is risking its own life, probably the lives of its own progeny that would not survive without their parent, and the possibility of subsequent progeny of its own, but there does not appear to be any compensating benefit in a Darwinian sense.

From a Dawkins "gene's eye view" there is a workable tradeoff. Members of the same species as our subject animal have perhaps 99.9 percent of the same genes. Animals of a particular species that live in the same area and are in the same herd or other similar local grouping are obviously likely to be much more related. Maybe they have 99.99 percent the same genetic data. So in the gene view, an animal protecting another such animal is actually increasing the chance "its" genes will be propagated. Remember that progeny in any event only express 50 percent of an animal's "personal" genes. From a gene point of view there would not appear to be a big difference between protecting one's own progeny and protecting the progeny of an animal that had 99.99 percent the same genetic data.

Discoveries in genetics suggesting that evolution of genetic data is a long-term process support the selfish gene theory.

We can summarize the properties of the Selfish Gene Theory of Evolution (SGT) as follows:

-SGT deemphasizes the importance of survival and reproduction. A characteristic can evolve if it aids in gene propagation even if it does not improve survival or reproduction.

-SGT deemphasizes the importance of the individual and "individual fitness."

-SGT is an extension or adjustment to classical Darwinian evolution theory. Because it is an extension, it encloses Darwinian theory and continues to explain all the observations explained by Darwinian theory. Dawkins takes great pains to say that his approach is only Darwin's theory from another view. "The selfish gene theory is Darwin's theory expressed in a way that Darwin did not choose but whose aptness, I should like to think, he would instantly have recognized and delighted in." Despite this disclaimer, Dawkins' view actually is a significant modification to classical Darwinism. Believers in orthodox Darwinism (and there still are many) are unlikely to believe in the selfish gene theory.

-SGT is based on post-Darwin discoveries regarding the actual mechanics of gene propagation. Obviously, if evolution occurs because of the propagation of genes, then the details of said propagation including chromosomes, meiosis, gene crossover and genetic distance are crucial to any theory of evolution, a point studiously ignored by believers in orthodox Darwinism. Darwin, knowing what he did *then*, certainly would *not* have "delighted in" the selfish gene theory. Darwin, knowing what we do *now*, may well have "delighted in" the selfish gene theory. There is a lot of research still to be done regarding gene propagation mechanics. Specifically, the ongoing research on human variations and single nucleotide polymorphisms will add a lot to this understanding.

So does the selfish gene theory actually debunk the group selection theory? It certainly provides a plausible alternative in the case of altruism and intra-species aggression. Did publication of the SGT result in all the group selectionists silently folding their tents and creeping off into the gathering twilight of evolution theory obscurity? No. There are still unrepentant group selectionists.

In a wider sense, the selfish gene theory actually seems to support group selection and other extensions to Darwinism by proposing a plausible alternative to orthodox Darwinism. In effect, on one side of a great ideological divide, we have the orthodox Darwinists, and on the other, we have selfish gene theory proponents, group selectionists, evolvability theory supporters, believers in complex evolution processes, and anybody else that believes that at least some extension, relaxation, or modification to orthodox Darwinism is not only possible but also necessary. In addition, the selfish gene theory deemphasizes individual fitness which would appear to lend some support to group selection.

More specifically, some of the SGT arguments seem to provide some direct support to group selection. If individual organisms are "survival machines" dedicated to the propagation of genes and if genes live longer than species, then could we not also say that *species* are "survival mechanisms" dedicated to the propagation of those genes. It would appear that if this argument works for individuals then working for species would only be a matter of degree. "Survival of the fittest species" would therefore be a plausible evolutionary mechanism. Therefore, species-level group selection should work.

Dawkins did not propose SGT as an explanation for aging. Dawkins was personally acquainted with Medawar and favors (as of publication of the 1986 edition of *The Selfish Gene*) the mutation accumulation theory. SGT works for altruism because the additional risk to the protector is presumed to be slight and the benefit to the protected is presumed to be substantial, a seemingly reasonable assumption. Except in very special circumstances, suicide would appear to have maximum negative impact and minimal benefit. If you recall the discussion on the evolutionary value of life, traditional evolutionary theories of aging agree that the value of extended (post-reproductive-age) life is minor or (Medawar) even zero. Therefore a case can obviously be made that the gene-benefit of some group advantage outweighs the (zero or small) disadvantage of a limited life. We can therefore claim that SGT supports purposely programmed aging.

In a wider sense, SGT recognizes the need for an alternative to orthodox Darwinism and thus strengthens the case for other alternative evolution theories that also support programmed aging such as group selection and evolvability theory.

"Selfish gene" is a popular term and essentially a trademark of Richard Dawkins. However, a growing number of theorists are beginning to accept similar "longer-term", "population oriented" concepts regarding evolution.

Information Based Evolution Concepts

We have discussed in previous sections how inheritance involves the transmission of digital and coded *information* or "data" between parent organisms and their descendants. This data controls the inheritable *design* of the descendent organisms, is stored within the organisms, and is then passed in turn to their descendants. The information contains *functional* data. A mutation that alters this functional data, alters the phenotypic design of the descendent organisms. So far, there is no disagreement with orthodox Darwinism.

However, genetics discloses that the majority of the data being transmitted in the reproduction of more advanced organisms is *non-functional*. Changes to this data *do not* alter the functional (fitness) phenotypic design of the organism but *are* transmitted to descendants.

The genetic information also has a complex *organizational structure*, which itself has evolved. The digital transmission of genetic data required development of a *language* specifying the methods whereby information is coded and decoded. This language varies slightly between different organisms and may be almost as arbitrary as any other language. Organisms that evolved in a different part of the universe might have a substantially different genetic language.

More complex organisms have more complex mechanisms for processing and "handling" genetic information that are clearly affected by the organizational aspects and non-functional data. Where only changes to the functional data affect traditional fitness, *any* change to the exact, letter-by-letter sequence, potentially affects evolution. As organisms become more complex the functional data generally becomes an increasingly minor part of the total. More complex organisms tend to have relatively more intron data and other non-functional data.

The selfish gene theory specifically talks to the data structure known as a "gene", the conservation of genes between species, and therefore the long-term nature of genes. However, genes, per se, are not the only data structures with obvious implications for the process of evolution. Repeats such as alu elements, introns, and other structural aspects also influence the evolutionary process.

8. Evolvability Theories

Evolvability theory refers to the idea that organism populations and species can vary in their capacity for evolution and that this factor needs to be considered and handled by any theory of evolutionary mechanics. In this context evolvability is essentially another proposed modification or adjustment to traditional evolution theory like group selection, kin selection, or the selfish gene theory. Evolvability considerations provide explanations for all of the discrepancies with traditional evolution theory including programmed aging, other lifespan limiting design features (biological suicide), sexual reproduction, elaborate evolved inheritance mechanisms, altruism, mating rituals, and excess male puberty age.

Evolution of Evolvability

It is generally accepted that organisms possess design features that enable the process of evolution. For example, all organisms possess the ability to pass information describing their designs to descendants, to store that information during the life of the organism, and to copy the information for distribution to multiple descendants, in addition to mechanisms that support accumulative adaptive modification of that information. The question here is whether it is possible for design properties that support or enhance the evolution process to *vary* between different organisms. If such was possible, then could not organisms evolve improvements in their ability to evolve? Would not such enhancements represent an obvious benefit in that organisms possessing them would be able to adapt more rapidly or comprehensively to changes in their environments? Would not any theory of evolution need to deal with variation in the capacity of organisms to evolve?

Traditional evolutionary mechanics theory ignores the evolvability issue. Either of two assumptions supports such a position. The first is that the capacity for evolution is a fundamental property of life that does not and cannot vary between populations or species, and that therefore evolvability is a constant that does not need to be considered in devising theories of evolutionary mechanics. The second is that evolvability is enclosed in traditional concepts of fitness and is therefore covered by traditional theory. Arguments are presented below to the effect that neither of these assumptions is correct.

Darwin's mechanics concept does not consider the capacity for evolution to be a variable. Darwin evidently and quite reasonably assumed that all living organisms had the capacity for evolution, that is, the capability for adapting by means of natural selection to changes in their external world of predators, food, habitat, etc. He also must have assumed that this capacity was a constant and not affected by other characteristics of organisms that might arise in the course of evolution. We can state these assumptions as follows:

-Darwin's theory assumes that all organisms possess, as a fundamental property of life, the capacity for adaptation, that is, the ability to evolve or evolutionary capacity.

-Darwin's theory further assumes that this fundamental property of evolutionary capacity is a constant and cannot be affected, negatively or positively, by any evolved characteristic.

-Therefore, Darwin's theory assumes that all living organisms have the same amount of evolutionary capacity and are therefore able to evolve, that is, adapt to external conditions, at the same rate. No organism is any better at adapting than any other organism.

If all three statements are not true, then an organism could evolve characteristics that increased its ability to evolve. It could therefore adapt to its external conditions more rapidly than some competing species. This would be an obvious competitive advantage that is not handled by Darwin's mechanics.

Most people would readily agree that the first statement is true: all living organisms have the properties necessary for evolution. However, what are these properties? Is it really true that all properties that contribute to the ability to evolve are fundamental properties common to all organisms?

As suggested in previous sections, subsequent developments, including essentially the entire science of genetics (unknown to Darwin), have provided substantial evidence disproving these assumptions regarding evolutionary capacity (or ***evolvability***). In fact, evolved traits of organisms *can* contribute to or detract from evolvability. The issues of evolvability have, since about 1994, resulted in development of a whole branch of theory. A *PubMed* search for "evolvability" in November 2008 returned 247 journal articles (333 in July 2010 and 495 in January 2013). Evolvability has applications outside of biology such as the development of computer programs capable of progressive and cumulative adaptation.

Evolvability appears to be a very long-term issue. Species such as the cockroach and fern have apparently survived for a very long time without significant evolution. However, evolution itself is a long-term issue in that same sense.

Evolvability is substantially different from and in conflict with traditional individual fitness. Many, possibly all, design features that *increase* evolvability *decrease* fitness or have no effect on fitness.

Evolvability theory as defined here is the idea that a design feature that resulted in an increase in evolvability could evolve despite an individual fitness disadvantage. The design of an organism is a compromise between fitness and evolvability.

A belief that the evolution process is the same in all organisms appears to require one to ignore much of the science of genetics. However, a belief in evolvability appears to be incompatible with orthodox Darwinism in that evolvability features are adverse or neutral regarding individual benefit. Darwin's dilemma redoux!

Possession of the *capacity* for evolution does not necessarily mean that a species *will* evolve. Evolution is driven by the need to adapt to some change in an organism's external world.

The reader has no doubt by now correctly guessed that this is leading to an *evolvability theory of aging[37]*. Weismann's 1882 theory may be the first evolvability theory of aging.

One property that Darwin put forth as *required* by his theory of natural selection was genetically controlled natural *variation* in traits possessed by organisms in a population. While Darwin presumably thought that such local variation was a fundamental property of life itself, as it would be if inheritance was an analog process, we now know (see Digital Genetics) that complex evolved adaptations such as meiosis, gene crossover, paired chromosomes, and X inactivation support variation due to recombination in the actual digital inheritance scheme. Organisms that sexually reproduce evolved from organisms that did not and did not have the degree of and quality of local variation possessed by the more advanced organisms. Reproductive techniques of primitive organisms tend to be more nearly like cloning or simple copying of genetic code that do not involve the complex shuffling procedures associated with sexual reproduction and therefore do not deliver the degree of variation. It is actually far easier for nature to make an identical copy of the digital genetic instructions than to produce structured, organized, and genetically transmittable variation.

It is therefore apparent that evolvability is affected by evolved traits at least in this one case of recombination. It is also apparent in this case that an *increase* in evolvability resulted in a *decrease* in individual fitness as suggested earlier. The logic for this is as follows:

In a hypothetical fully adapted organism that was totally optimized by natural selection to its external world, every parameter would be optimum or at least as optimum as it was going to get. For example, if a certain height (to name one trait of an animal) were optimum, then we would expect that the average height of all the animals in the population would be the optimum height. However, recombination would cause some of the animals to be shorter than optimum and some to be taller than optimum. This is the familiar "bell shaped curve." All the shorter or taller animals are less than optimally fit. The more variation there is, the less fit the average animal will be and therefore the less fit the population as a whole will be. A population with more variation would be less able to compete with a rival population that had less variation.

Nevertheless, variation enables natural selection. The more variation that exists the greater the capacity for adapting to *changes* in the animal's external world. Variation *presets* a situation in which some of the animals will be instantaneously better adapted than others with regard to a *change* in the world. If the entire population consisted of genetically identical clones of an optimum animal, then all the animals would be the same optimum height and the average fitness of the population would be maximized. However, at the same time, evolvability for this population would be zero. These animals would be completely incapable of adapting through natural selection to any changes in their world because there are no genetically transmittable differences between them for natural selection to select.

Orthodox Darwinists would say that the occasional mutation would introduce changes and that therefore variation would never be zero. However, as indicated earlier, recombination creates much larger variation because of cascading the effects of individual mutational differences.

So here we have a case of an *evolved* trait (the variation producing aspects of sexual reproduction and meiosis), which *reduces* fitness. Animals possessing this trait are *less likely* to survive and breed than animals not possessing the trait, a violation of the rules for Darwinian natural selection! However, animals *must* possess this trait in order to evolve. Although we cannot say that all *current* species have evolvability, non-zero evolvability must have existed in all of their *ancestor* species. It is therefore clear that a tradeoff must exist between traits resulting in local variation, and fitness. An ancestor species that had inadequate fitness would have died out and left no progeny. An ancestor that had inadequate evolvability would also have died out because it was unable to adapt.

A second discrepancy between orthodox Darwinism and modern genetics involves the class of design changes that change genome organization or modify "junk" DNA in such a way that the ability of an organism to adapt is altered without affecting fitness. This class of design changes fits with evolvability theory.

The inheritance system and existence of a complex, obviously evolved, genomic design in sexually reproducing organisms generally represents a conflict with the idea that natural selection, selecting between phenotypic differences in individual organisms, completely explains evolution. To illustrate, a text document (also digital data) could be written completely defining the design of some complex structure. The document could then be copied and distributed to multiple builders for execution. The methods and systems used to copy and transmit the data do not affect the design of the structure; (the designer could even have written in a different language). In the same way genomic design, which also involves a language, decoding, interpreting, merging, copying, and other complex processes does not affect the phenotypic design of organisms defined by the data it carries. Therefore, individual benefit cannot explain the development, evolution, and retention of the sexual inheritance system. What drove evolution of the genome?

The need for evolvability depends on evolutionary pressure. A species that, for whatever reason, did not encounter many changes in its external world would not need much evolvability. However, higher animals would have a great need for evolvability because, if for no other reason, other organisms were evolving. An animal whose predators or prey were evolving would have great need for evolution itself. The *rate* at which a species could evolve could well determine its fate. Regardless of how one defines "species" in a time-sequential context, it is obvious that very many species occupy the chain of descendancy between the original single-cell primordial life form and any current animal.

As you will see, there are many other characteristics of an animal that could contribute to evolvability.

Evolvability, like the selfish gene theory, is based on discoveries in genetics that were made subsequent to Darwin. Like the selfish gene theory and group selection theory, evolvability is an extension or adjustment to orthodox Darwinism. Like selfish gene theory, we could presume that, if Darwin were here today and knew what we do about genetics, he would "embrace" the evolvability theory. Both are incompatible with orthodox Darwinism. Finally, evolvability, group selection, and selfish gene theory appear to be compatible and are not mutually exclusive.

Death Rate and Evolvability

A requirement of the theory of natural selection is that evolution requires deaths. Evolution results from the differences in average lifespan between more fit and less fit organisms. A hypothetical organism that did not die could not evolve.

In effect each life of an organism is a trial of and a "vote" for the combination of traits possessed by that organism. A longer than average life and thereby production of more progeny represents a vote in favor of the combination of traits exhibited by that organism. A shorter life is a vote against.

It should therefore be apparent that *death rate* is a factor in evolvability. A higher death rate will accumulate more votes and test more combinations more rapidly.

As larger, more complex, animals evolved, they encountered reductions in evolvability for several reasons:

First, larger animals consume more resources and their populations therefore tend to be smaller (how many elephants are there in the world compared to ants). Smaller populations have lower death rates. A smaller population represents a smaller number of simultaneous combinations of traits and a lower rate at which trials are being conducted.

Second, larger and more complex animals tend to require more time to develop into mature adults. Their lifespans must be longer to accommodate the longer development time. As we shall see, deaths of animals prior to becoming mature do not contribute to evolution in the same way as deaths of mature organisms. Therefore, even for the same size population, longer living animals will have lower death rates and therefore have an evolutionary disadvantage.

Third, more complex organisms have more combinations of characteristics to be sorted out. More simultaneous equations need to be solved. Are longer claws and shorter feet better, or are longer feet and shorter claws better, or are longer feet and longer claws but a larger, heavier, and therefore slower, animal better, and so forth for thousands of parameters. More complex animals presumably need more "votes" to evolve.

In effect, the larger and more complex an animal becomes, the more difficult further evolution tends to become *unless compensating factors are present*. Evolvability theory proposes that more complex organisms have evolved many ways to increase their evolvability and that evolvability in such organisms would be negligible without these enhancements.

If death rate is a factor in evolvability, it should be apparent that animals that had an unnecessarily long lifespan would be at a disadvantage with regard to evolvability. There will therefore presumably be a tradeoff between the fitness advantage of a longer lifespan and the evolvability disadvantage of a longer lifespan. This would appear to be a reasonable explanation for aging or other lifespan control mechanism as an evolved trait. Further analysis has disclosed several other reasons that aging could represent an evolvability benefit.

Adult Death Rate

Another implied requirement of Darwin's theory is that in order to be selected, a characteristic has to be expressed in such a way that that it affects the differential in lifespan between organisms that have the trait and organisms that do not or otherwise affects the organism's ability to propagate its design. A characteristic that is latent at the time an organism dies cannot have influenced the probability of propagation and therefore cannot have contributed to the evolution of that characteristic.

Most external survival traits of animals such as speed, intelligence, strength, size, coloration, hunting ability, ability to withstand the environment, and so forth, *are not fully expressed* until the animal is a *mature adult*. Therefore, in order to contribute to evolution of these traits, a death has to occur when an animal is a mature adult. Deaths of juveniles do not contribute to evolution of traits that are not fully expressed in juveniles. Therefore, traits that contribute to increasing ***adult death rate*** increase evolvability. We can consider adult death rate as the rate at which nature conducts tests of different combinations of mature inheritable traits. The more tests, the faster evolution can proceed.

Protection of Young

Animals that protect and nurture their young evolved from animals that did not. An animal that protects its young has a larger chance of itself dying and not having subsequent progeny than one that does not, a fitness disadvantage. However, the chance of progeny surviving is presumably enhanced, a fitness advantage.

From an evolvability standpoint, protection of young has a great impact on effective adult death rate as follows:

If the adult young-protecting parents of an immature animal die, the progeny in a wild situation will almost certainly die. This, in effect, shifts deaths of some juveniles to the "adult" category because the death of a protecting, nurturing adult will very likely result in the death of its immature progeny. In other words, if an immature animal dies, it is rather likely that the characteristics of its adult parents had as much or more to do with its death as its own characteristics. In this case, the deaths of the immature animals counted, for evolutionary purposes, as adult deaths. They died because their parent(s) died. Protection of young increases "effective" adult death rate and therefore improves evolvability in mammals and other more advanced animals.

The Cycle of Life

There is a very simple birth-death equation that applies to populations of living things:

Average Birth Rate = Average Death Rate

If long-term average birth rate for a species exceeds average death rate by even a little bit, then in a significant time period (say 5,000 years) the planet would be completely covered by organisms.

If average death rate exceeds birth rate then the species becomes extinct.

Of course, there are famines, floods, blooms, overpopulation events and other conditions that cause populations of any species to increase and decrease on a temporary, short-term basis. Successful organisms are able to accommodate these short-term situations. More forcefully, the capacity for coping with short-term events such as these is a survival trait that would tend to be selected. As an illustration, suppose a famine or drought caused a major decimation of the populations of small animals in a region. After the event ended a species that could reproduce rapidly would have an advantage and might be able to take over territory from other, more slowly reproducing, species.

Darwin considered that the populations of all organisms were controlled entirely by *external* "checks" such as predators, food supply, disease, and environmental conditions. However, it is obvious that more complex animals have *internal*, genetically programmed, features that act as restraints on breeding such as the following:

Age at puberty limits breeding to those animals older than a genetically set age.

Length and frequency of fertility periods limits breeding to those periods. Many animals are only fertile at certain times of year, often only once per year. The timing of the fertility period, often in the fall, helps the survival of young that are typically born or hatched in the spring.

Gestation periods limit reproduction in most animals in the sense that the animal can only become pregnant once during the gestation period.

Litter size limits the number of young produced per pregnancy.

Mating rituals often prevent younger, weaker, and smaller animals as well as older, weaker animals from breeding thus limiting reproduction.

Aging tends to limit breeding to younger animals because of reductions in breeding vigor.

Other Behavior Patterns such as societal restrictions can limit reproduction.

Traditional biologists would say that all these traits (except aging and mating rituals or other behaviors) could be fully explained in purely fitness terms without invoking evolvability. Gestation periods are what they have to be in order to accomplish the required tasks. Puberty is what it has to be because younger and smaller females could not accommodate gestation and birth. Fertility periods have to be in the fall to avoid infant mortality associated with bearing young in the winter. All of these arguments are valid.

However, it is clear that these *life cycle characteristics* can also have an evolvability purpose.

If puberty age were older, then fitness would be reduced. An animal would have a greater chance of dying prior to breeding. Its chances for producing progeny would be reduced.

However, evolvability would be increased because animals that survived the longer period would presumably be more fit, and therefore their progeny would be more fit. In addition, an older puberty age would contribute to evolvability via increasing adult death rate. The probability would be increased that progeny were a result of animals that fully exhibited adult characteristics.

Therefore, we see that puberty age *could* represent a tradeoff between evolvability and fitness.

It is interesting to note that there does not appear to be any traditional Darwinian reason for *male* puberty age to be as old as it is in many animals. Physically, males could accommodate much younger puberty. In humans, various disorders cause puberty to occur as young as 4 years of age. A male with a younger puberty age would appear to be more fit because it would be likely to have more progeny. In mammals and other organisms that protect and nurture young, an argument could be made that a reduced male age at puberty would represent an individual disadvantage because the animal would be less able to perform the protection function. However, turtles lay their eggs and walk away so there is no protection and nurturing function. Why then do male turtles have such a delayed age at puberty?

The problem with the idea that populations are controlled only by external checks is that it does not seem to be very efficient. A species that was not controlled by predators would tend to breed itself into overpopulation and subsequently be in a state of starvation and probably decimated by disease. Populations would constantly cycle up and down. In a down cycle territory could be annexed by other species.

While populations of simple organisms may in fact be controlled exclusively by external checks, would it not make sense for more complex animals to evolve some internal methods for controlling population to avoid these consequences?

Populations of animals do in fact experience these types of cycles but not to the extent that one would expect. Are back yard squirrels usually starving? Do they seem to be constantly diseased? Do predators constantly chase them?

If you accept the idea of evolvability and the issue of adult death rate as described above, there is another problem with population being exclusively controlled by external checks. If a population of animals is controlled by starvation, adult death rate will tend to be adversely affected. If every animal breeds when it reaches puberty (as proposed by traditional aging theory models), then the only way to keep the birth-death equation balanced is if most animals die prior to puberty due to starvation, infant mortality, disease, etc. This obviously adversely affects adult death rate and thereby evolvability. The *species* (or at least more evolved descendent species) would be better off if it could somehow reduce birth rate so that more individuals lived to be adults.

Looking again at the birth-death equation, births are controlled by the combined effect of all the life-cycle characteristics, and deaths are controlled by external checks and the interaction of aging with external checks. If a species is in fact capable of (to some extent) controlling its own population to maximize adult death rate and otherwise optimize its situation as described above, then such control must be by means of varying one or more of the life-cycle traits.

Of course, any trait leading to such control is adverse to individual fitness and therefore incompatible with traditional natural selection.

There is a lot of evidence that some animals in fact can control some of their life-cycle characteristics.

Let's look a little closer at how the life cycle characteristics interact. Consider a hypothetical stable population of mammals in which we could somehow adjust life cycle characteristics. If we were to reduce age of puberty, animals would begin reproducing at a younger age than before, and the birthrate would increase. Unless there was a compensating change in some other internal characteristic such as more aggressive aging, or reduced average litter size, median lifespan would have to decrease to balance the birth-death equation.

Similarly, suppose we reduce the aggressiveness of the aging characteristic. Now some animals are living longer and having more progeny than before. Again, unless there was a compensating change such as an increase in puberty age or a more restricting mating ritual, the *median* lifespan of the population would have to *decrease* in order to keep the birth-death equation balanced. If some animals are living longer and therefore reproducing more, other animals must be living shorter lives and reproducing less.

Either of these outcomes is undesirable from an evolvability standpoint because they reduce adult death rate.

This concept seems to fit observed animal characteristics. Animals with very long lifespans (e.g. Sturgeon) also take a long time to reach sexual maturity.

To summarize, orthodox Darwinism maintains that all organisms are trying to live as long as they can and reproduce as much as they can. All limitations to life-span or reproduction are either externally imposed (predators, food supply, etc.), fundamental limitations (physics, chemistry, etc.), or the result of a trade-off with a beneficial function. Variation is a fundamental property of life resulting from occasional propagatable mutations.

In the evolvability view, organisms are trying to conduct as many trials or tests as possible, where each test is the result of an adult life. Organisms are also trying to maintain variation, which is not a fundamental property of life. This is important because each adult in a varying population represents a different combination of characteristics and therefore supports the evolutionary process. Lifespan and reproductive characteristics may be adjusted to maximize the rate at which these tests are conducted, i.e. adult death rate.

Mating Rituals

Mating rituals have long puzzled biologists.

Many behaviors of animals are perfectly consistent with fitness. An animal fighting peers for food is consistent with fitness. Stronger, smarter, faster animals will prevail and survive and pass their genes to progeny. Similarly, an animal fighting for territory or fighting to mate makes sense from a fitness point of view.

However, some behaviors seem adverse to fitness. An animal should mate with the first mate it finds rather than wait for a better mate because otherwise it might die before mating. Any kind of selectivity seems to represent a reduction in the probability that an animal would breed and therefore a reduction in fitness. So how did such a trait evolve? An animal possessing such a selectivity trait (waiting for a better mate) would be less likely to survive long enough to mate and therefore less likely to pass its genes into the pool.

Many mating rituals seem designed to select animals for mating based on some aspect of quality such as strength, or speed, again inconsistent with fitness but consistent with evolvability.

Many mating rituals seem to have the general effect of delaying mating until animals are more mature which is consistent with evolvability but inconsistent with fitness.

Some traditional biologists dismiss even elaborate and structured mating rituals as merely instances of competition over mates and therefore completely explainable in fitness terms. Mating rituals disturb others.

Mating of the Bighorn

As one of the most spectacular examples of a mating ritual, consider the Bighorn Sheep (*Ovis canadensis*) that live in the Rocky Mountains of North America. Bighorn reach sexual maturity in two years, mate only in November and December, have a gestation period of 6 months, bear one or two young, and live about 15 years in the wild.

The Bighorn have evolved an instinct (let's call it the head-butting instinct) that leads them to have head-butting contests to determine which rams are to mate with the females. (Females have also been observed in head-butting contests.) Such contests have been known to last as long as 24 hours. To support the head-butting mating ritual, the Bighorn have evolved extremely large and heavy horns that weigh as much as 10 percent of the entire animal's weight. Increased skull size, spine, and muscle mass needed to support the horns probably represent another 10 percent.

Although sexually mature at 2 years, the average male does not mate until 7 years of age because the mating ritual requires animals to be older and stronger to mate.

The head-butting mating ritual instinct has a negative effect on individual fitness since an animal that had the instinct would be less likely to breed than one that did not. The probability of an animal breeding is severely reduced by the mating ritual because it has to survive on average for an additional *five years* (after sexual maturity) in order to breed, and because it has to pass the "test" imposed by the mating ritual. (Some sheep never get to mate.)

Figure 24 Bighorn Sheep

In addition, the excess size of the horns is apparently a significant disadvantage to the Bighorn as it relates to its world of food, predators, and environment. The horns have little value in resisting predators and have no value in helping to obtain food. The extra weight of the horns is a disadvantage for both. The head-butting contests are noisy and attract predators.

However, the mating ritual promotes the evolution of beneficial characteristics as follows: Presumably, the test imposed by the mating ritual selects animals with desirable characteristics such as strength, stamina, and agility. In addition, by delaying mating until animals are older and stronger, the mating ritual allows generic natural selection more time to work. Animals will have to pass a longer "life test" to breed. Less competitive animals have a greater chance of having died prior to breeding.

The Bighorn illustrate a major error in the models used by traditional theorists Medawar and Williams, namely, actual animals often do *not* start breeding at puberty. As we shall see, if the characteristics displayed by actual animals are considered, aging has a much more profound effect on populations than predicted by the simple model.

The Bighorn have existed for millions of years under a regime in which stronger animals had preference in breeding. If we removed the mating ritual from the sheep (as might be done experimentally by using forced random breeding in captivity) we would expect the strength and other "beneficial" qualities of an average sheep to decline very rapidly, maybe in the first few generations.

The Bighorn mating ritual also illustrates what appears to be a population sensitive control on breeding. In an area of low population, the mating ritual could result in little delay and therefore reduced impediment to breeding because there would be fewer animals to compete.

Maybe breeding could begin at three years. (In a limit case where only one male and one female were present, the ritual would have no delaying effect.) In an area with a higher population density, competition would be greater resulting in a larger average delay. This population sensitive feature aids the species in rapidly repopulating after an event such as a famine while concentrating on quality and evolvability once a substantial population has been achieved.

The Challenge Effect

The mating ritual described above seems to have a *challenge effect* in that animals have to pass a test in order to breed. In areas with higher populations of animals, only older and stronger animals can pass the test and breed. An animal that had some beneficial trait such as larger size, or increased strength might well be able to pass the challenge and breed a year *earlier* (younger) than typical. An inferior animal might only be able to breed *later* (older) than typical or might be totally unable to breed.

Aging has a challenge effect very similar to that of the mating ritual. Also, aging *interacts* with the mating ritual. As animals become older, they also get weaker, slower, less agile, and less able to pass the challenge. However, an exceptional animal, possessing a beneficial trait such as strength or size might well (as suggested by Skulachev below) pass the mating ritual *despite* the effect of aging. An animal with desirable traits might be able to begin breeding *a year younger* than average and also continue to breed *a year older* than average.

Vladimir Skulachev is director of the Belozersky Institute of Physico-Chemical Biology at Moscow State University and an Academician in the Russian Academy of Sciences.

Skulachev believes that aging *is* an extension of programmed cell death as described in his 1997 paper[38] titled *Aging is a Specific Biological Function Rather than the Result of a Disorder in Complex Living Systems: Biochemical Evidence in Support of Weismann's Hypothesis*. Specifically, analysis of the three main mechanisms proposed as proximate causes of aging (telomere shortening, heat shock proteins, and oxidation. See Aging Mechanisms.) indicated to him that it was extremely unlikely that these mechanisms were accidental disorders.

Skulachev proposed that the function of aging was as originally proposed by Weismann, "to reduce the pollution of the population by long living ancestors thereby stimulating evolution." However, he also proposed that the gradually increasing deterioration caused by aging could serve evolution in an additional way:

> "The appearance of a useful trait allows compensation of the effect of aging within certain time limits. A large-bodied deer, even after reaching an old age, has better chances to win a spring battle for a female or escape from a group of wolves in comparison to a younger but smaller conspecific animal."

In other words, an animal having a sufficient fitness advantage could survive and breed *despite the gradually increasing deleterious effects of aging*.

Decline in sexual vigor is also a challenge. An older Bighorn, even though stronger, might decide that mating was not worth a 24-hour battle accompanied by a massive headache. A yet stronger animal might decide that it was despite reduced urge to mate.

Note that in Bighorn, as in many animals, mating opportunities occur only annually. A female, once impregnated, cannot be further impregnated until the next mating season.

Therefore the *length* of an animal's actual *breeding period* (number of years that it was reproductively active) is a critical factor in determining the number of progeny produced by that animal. This period is constrained at the younger end by puberty and the mating ritual, and is constrained on the older end by either death of the animal or the effects of age on strength, or other applicable trait, and/or strength of reproductive urge. The mating ritual interacts at both ends of the breeding period.

Even if there was no mating ritual, aging provides a challenge mechanism aiding in the selection of beneficial traits. An animal having beneficial traits is more likely to survive longer and thereby have a longer breeding period despite the weakening effects of aging than an inferior animal.

Evolutionary Disadvantages of Immortality

Perhaps the best way to describe why aging is a necessary evolved adaptation is to consider the many evolvability disadvantages that would be encountered by a non-aging species.

Challenge Effect: Animals without an aging mechanism would not possess the challenge effect that aging provides in helping select beneficial characteristics as described in the previous section. Aging would not constrain the older end of an animal's breeding period.

Adverse Effect of Experience: Actual animals, especially more advanced animals, have a capability for learning from experience. Even worms have some learning ability. Experience will make animals more capable of dealing with their external world of predators, prey, food supply, and environment. Because of this, an older, more experienced, non-aging animal will be able to out-survive and out-breed a genetically superior but less experienced younger animal, an obviously bad outcome from an evolutionary standpoint.

The probability of an older, non-aging animal dying in any given time period is therefore lower than that of a younger, physically and genetically identical, mature, non-aging animal. Given mating rituals, competition for breeding and so forth, it is obvious that an older non-aging animal could also have more progeny in any given interval than a genetically superior younger animal.

A very similar aspect is "pecking order." Since more advanced animals have memories they can remember their position and the positions of other animals in the pecking order that determines mating rights and other privileges in a typical group of animals. Once an animal achieves some position in the pecking order, it is likely to be able to maintain that position for at least some period of time without further competition. A non-aging animal might be able to maintain such a position in such a way that genetic merit is contravened.

Death rates for animals capable of learning will therefore *decline* with age instead of being constant as a function of age as stated for the traditional aging theories. Breeding rates for non-aging animals will *increase* with age instead of being constant. Because of the experience factor and memory factor, the fitness impact of aging is greater in actual animals than suggested by traditional theorists. In effect, an acquired non-genetic trait (experience) is competing with genetic traits for selection, an evolvability disadvantage.

Adverse Effect of Immunity: A very similar situation exists with regard to immunity from infectious diseases. If an animal is exposed to an infectious disease, it has some probability of dying from the disease (or from predators or environment or starvation as a result of the weakness resulting from the disease). If it survives, it obtains some immunity against subsequently contracting the same disease. Subsequently, that animal has a lower total chance of dying because of immunity to at least that one disease. Because of this immunity factor, non-aging animals have a further declining probability of death, as they get older.

Adult Death Rate: As described above, even surviving aging animals have a declining breeding rate with age because of declining motivation. In a non-aging population, especially in view of the effects of experience and immunity, older animals would produce far more progeny than in an aging population. Because of the birth-death equation, this means that a far larger proportion of the animals would have to die without producing progeny, which is adverse to genetic diversity and adult death rate.

Hypothetical Case: Consider some hypothetical non-aging animals. These animals have a group structure in which the dominant male mates with all the females and the other males do not mate. The females do the work of finding and gathering food, as well as protecting, and nurturing the young. This work represents most of the hazardous activity. The dominant male is somewhat protected and served by the other members of the group. If a young male exhibits traits that indicate that it might someday be a serious threat to the dominant male, it is killed or forced out of the group by the dominant male.

Now consider the tremendous negative evolutionary effect "absence of aging" would have for these animals. Because of the protection, experience factor, immunity factor, and group dynamics, the dominant male could expect to live for a very, very long time and sire all of the group's progeny during that period. He could be mating with many generations of his own descendants. Genetic diversity would be a joke. An animal that could have gained its dominant male position through luck as opposed to genetic merit can maintain it using acquired non-genetic factors.

Notice the dramatic difference between this case and the traditional (Medawar) model for non-aging animals. These hypothetical animals more resemble actual animals like lions, gorillas, and (probably) primitive humans than the traditional model.

Mechanics of Evolvability

Recall that the absence of a mechanism for the propagation and retention of an individually adverse trait is one of the major traditional objections to adaptive theories of aging. The following is an attempt to show one way in which such a mechanism could work for evolvability.

Imagine a population of some animal species that are in a changing environment and therefore under evolutionary pressure. The average fitness of animals in this population is "F". Now imagine that an evolutionarily short time period passes. We could even consider a period as short as one month. During this period some animals would die. Presumably, statistically more of the less fit animals die. During the period some animals are born. These new animals are statistically more likely to be the descendants of more fit parents and are therefore more fit. Therefore at the end of the period, the average fitness of animals in the population has increased by some small amount ΔF.

Now imagine a second population that is nominally identical to the first and has the same average fitness. However this population has a lower evolvability. Perhaps the variation between individual animals is less. At the end of the period the average fitness of this second population would have increased, but by an amount smaller than ΔF.

The message here is that evolvability is not a long-term issue. Evolvability operates on the same time-frame as evolution itself.

Let's consider another group "A" of hypothetical mammals. These animals have a mating ritual that tends to restrict mating to animals that are more fit using mechanics similar to those described earlier in the discussion on mating rituals. Another group "B" is identical to "A" except they do not have the mating ritual. Since the mating ritual restricts breeding it represents an individual fitness disadvantage for the "A" animals.

Now consider the next generation consisting of animals resulting from unions between animals in the original groups. Because of the natural variations in the animals and because the animals in the next generation of group "A" are the result of unions between animals that are generally more fit, the average animal in the next generation of group "A" must be more fit than the animals in the next generation of group "B." Because they are more fit they are, by definition, more able to propagate their traits including the mating ritual.

Aging, for reasons already enumerated (the challenge effect, etc.), acts in ways very similar to such a mating ritual to increase fitness in the next generation.

The mating ritual and aging serve to amplify natural selection by increasing the breeding advantage of animals that are more fit and increasing the breeding disadvantage of animals that are less fit.

The tradeoff between individual fitness disadvantage and "next generation" fitness advantage involved here does not appear to be very different from the tradeoffs between survivability and reproductive ability, or between survivability, reproductive ability, and ability to protect young, which are generally accepted. However, traditional theorists and group selectionists would point out that there is a significant difference in that the traditional fitness tradeoffs result in a net benefit to an individual while aging and mating rituals result in a net individual fitness disadvantage and only have a positive "collective" advantage. For example, protection-of-young, while a disadvantage for the parent is an advantage for the progeny of that particular parent and increases the chance of that parent propagating its individual genes, an individual advantage. Mating rituals, other instances of sexual "selection", and aging, are not individually beneficial.

Are mating rituals and aging therefore instances of group selection as suggested by group selectionists?

Evolution itself, at least "Darwinian" evolution by means of tiny incremental steps, requires a population. Random chance is much more important to the fate of an individual animal than any such tiny increment in fitness. Normal variation in animal traits is much larger than a tiny increment. Evolution requires a population large enough and a time long enough to "average out" the effects of chance and variation and resolve the effects of the incremental improvement.

The mechanism suggested above therefore does not appear to require a "group" of a size larger than that required for generic natural selection. Also, the effect of such an amplifying trait is very immediate, (one generation) and therefore the benefit is not delayed from or slower than the effect of the individual disadvantage, a perceived problem with group selection.

Let us consider another pair of hypothetical evolving animal populations, C and D. The Ds are initially identical to the Cs except they have a longer development time and have a lifespan twice as long. According to Darwin's theory of incremental evolution, each generation is minutely more adapted than the previous generation. Each generation thus accumulates an increment of additional fitness we could call "dF." If we look at the Ds in five generations, they have accumulated 5 dF of additional fitness. During the same period the ten generations of Cs accumulated 10 dF of fitness, an obvious advantage.

This assumes that the generational fitness increment is the same between the Cs and Ds. Is it not possible that the longer lifespan of the Ds confers a greater fitness increment in each generation? If the Ds live longer, then natural selection has a longer time to work and the Ds that survive to breed in each generation should be more fit than the corresponding Cs. This is a valid argument. However, it is also clear that there must be a point of "diminishing return." The same arguments that state that there is a "decline in the fitness effect of adverse events with age" would apply to the diminishing benefit of a longer lifespan. There must therefore be an optimum lifespan relative to development time.

Note again that it does not take millions of years or the extinction of a species for an evolvability advantage such as shorter lifespan to take effect.

Here is another way to visualize the idea that evolvability operates on the same time scale as natural selection: Evolvability characteristics *benefit the process of natural selection* by contributing to *preconditions* (such as local variation) needed for the operation of the natural selection process. We can imagine a relationship like:

dF/dt = kEP

In this equation, dF/dt is the rate at which fitness (F) would increase in response to evolutionary pressure (P) given a population evolvability (E). For a limit-case example, imagine a population consisting entirely of identical clones possessing identical genomes. We could assume for this exercise a species in which individuals can change sex and also assume that at the beginning of a time period our clones were perfectly adapted to the then current conditions. Evolvability in this population would be zero because there is no variation for natural selection to select, while average fitness would be maximized at the *beginning* of the period because all of the members of the population are perfectly adapted.

Now imagine a population in which there was more variation around the ideal design. Most of the members of this population would be less fit because they varied from the ideal but they would possess more evolvability and thereby ability to adapt to *changing* conditions.

The relationship suggested here is equally valid for any size time interval (dt) because (t) does not appear in the right side of the equation. We can therefore contend that evolvability is not subject to the sort of timing and sequence criticism directed at propagation of group selection characteristics.

As in the case of group selection, critics do not claim that organisms do not vary in regard to evolvability or that the specific evolvability benefits claimed for a limited lifespan are invalid. Their objection is to the propagation and retention of an individually adverse characteristic regardless of evolvability benefit.

Evolution of Intelligence

The evolution of intelligence appears to be qualitatively different from the evolution of other animal characteristics such as claws and fur.

Let us define *intelligence* as the genetically transmitted characteristics that allow information acquired during an animal's life to alter its behavior in a fitness beneficial manner. Intelligence presumably includes the ability to store acquired information (memory) and many complex traits that facilitate acquisition of information (such as curiosity) or processing of acquired information (such as associative ability).

We define *experience* as all of the animal's collected and stored (remembered) information concerning its world.

Wisdom is the beneficial combination of intelligence and experience. The probability of an animal surviving is affected by its wisdom. From a fitness viewpoint, intelligence without experience is useless. Intelligence is essentially the ability to learn from experience. Experience without intelligence is useless. Wisdom is the factor that would tend to be selected by natural selection.

This creates an interesting situation regarding the evolution of intelligence. As mentioned earlier, the evolution of any characteristic requires that the characteristic be expressed. This in turn requires the survival competition of mature organisms in which the characteristic is fully expressed. Following maturity, organisms would have to live long enough to allow a period during which the competitive advantage of some characteristic could become apparent. This is the period in which animals with superior characteristics live longer and breed more in order to make evolution work.

We can presume that intelligence is fully expressed when an animal reaches maturity. Intelligence is useless without experience so additional time would be required for animals to acquire experience so that their intelligence advantage would become apparent and thus allow selection of the more intelligent animals. Therefore, evolution of intelligence apparently requires a relatively longer lifespan than evolution of characteristics such as claws and teeth.

However, there appears to be another problem. Experience accumulates for the lifetime of the animal. An older and more experienced but less intelligent animal would be more fit than a younger, more intelligent but less experienced animal. An *acquired*, non-genetic characteristic (experience) is *competing* with the genetic characteristic (intelligence). If animals did not possess aging or some other lifespan regulating characteristic, would it be possible for intelligence to evolve? Apparently, lifespan regulation is required for the evolution of intelligence.

Immunity presents a very similar situation. Like wisdom, immunity involves the combination of evolved, very complex, *genetically transmitted* characteristics that provide the mechanics for the development of immunity, and *acquired* characteristics resulting from exposure to specific pathogens. Immunity tends to be cumulative. The longer an animal lives the more exposures it has and the more immunity it potentially acquires. An old animal is therefore less likely to die of an infectious disease than a younger, genetically superior animal. Here again, lifespan regulation appears to be required to support the evolution of the genetically controlled parts of the immunity system.

Some think of evolvability as equivalent to species-level group selection, that is, evolvability benefits the species. They inquire: How could a very long-term, deferred, large-group benefit outweigh an immediate individual disadvantage? Actually, evolvability benefits the ***evolution process***. The evolvability concept does not require a group larger than or a term longer than the fitness concept as illustrated in the previous discussion. More specifically, there is a major logical difference regarding the time-sequence of the cost-benefit tradeoff involved. In group selection, we are considering an individual disadvantage that is traded for a *future* group benefit. Despite the individual disadvantage, the design property must propagate to a sufficiently large group such that the group benefit is felt. The larger the group the more time must pass between cost and benefit, a significant issue regarding the feasibility of group selection. In the evolvability case, the design characteristics that produce evolvability must *preexist* in order for natural selection to work. The evolvability benefit is being traded against a future individual cost. No selection in either direction is possible in the absence of evolvability. Beneficial selection (selecting "in") or rejection of an adverse trait (selecting "out") are both reduced if evolvability is reduced.

The traditional approach to thinking about or analyzing the evolution process is to assume some selectable phenotypic difference and then track how natural selection operates upon that difference. This approach does not work for an evolvability characteristic because evolvability characteristics act to *create phenotypic difference* or to *enhance selection*. Therefore the traditional analysis captures the individual disadvantage of the evolvability characteristic but fails to account for its benefit, which occurred *prior* to the beginning of the analyzed sequence.

Evolvability provides an explanation for altruism and other individually adverse behaviors. Fitness is served by an animal mating with another animal as similar to itself as possible (e.g. a relative) in order that its descendants will be as similar as possible to the parent. Evolvability is served by an animal mating with one as different from itself as possible such that variation is maximized (e.g. non-relative of substantially different lineage). Evolvability theory proposes that actual animal behaviors are a compromise between fitness and evolvability. If it makes sense to mate with a non-relative does it not also make sense to protect a non-relative? The protectee might be a future mate, an ancestor of a future mate, or a future mate of a descendent of the protector.

In many animals sexual reproduction represents a massive individual disadvantage relative to simpler reproduction methods. In a population of turtles, the females lay a certain number of eggs per year. If they could lay twice as many eggs that would be an obvious individual benefit but, because of physical and resource limitations, the eggs would have to be smaller and less developed, an individual disadvantage. We presume the turtles have evolved a compromise between egg size and quantity. However, if all the turtles could lay eggs and not just the females, an obvious large individual benefit would prevail. Apparently sexual reproduction conveys such an advantage to evolvability as to outweigh the individual disadvantages.

Evolvability provides explanations (to compete with traditional explanations) for all the previously discussed discrepancies between traditional evolution theory and observations.

Evolvability Theory Summary

-Discoveries in genetics science during the last half of the 20th century are directly incompatible with traditional evolutionary mechanics theory and support the idea that populations and species vary in their capacity for evolution and can acquire design features that increase their ability to evolve.

-Many, possibly all design features that increase evolvability are individually adverse or at best fitness-neutral. Evolvability is a trade-off with individual benefit.

-Evolvability considerations provide explanations for all of the observed discrepancies with traditional evolution theory. Evolvability provides multiple explanations for design-limited lifespan.

The concept of evolvability introduces major complexity into thinking about the evolution process. Instead of a single simple evolution process we now have multiple different processes. Evolution in bacteria is not the same process as evolution in sexually reproducing species so that data acquired regarding bacteria evolution is not automatically applicable to complex organisms. Intelligent species evolve differently from non-intelligent species, and so forth. Evolvability properties vary from species to species in a manner similar to fitness properties.

9. Attitudes about Aging

Attitudes about aging are one of the non-science factors that influence scientific and public opinions about aging. Aging theory is very different from most other areas of scientific inquiry. For example, most people have little or no personally observed information about nuclear physics. Most people do not care that much about nuclear physics. As a result, if "science tells us" that matter is made up of atoms, and so forth, people have no reason to doubt "science" and little reason to have strong feelings about any particular scientific conclusion.

Aging presents an entirely different picture. The average person has extensive and very detailed information about human aging obtained from direct personal experience and observation. Aging is a major factor in the lives of most people. People care about aging. Aging has enormous economic, political, moral, and even religious impact. If "science tells us" this or that about aging, people have many reasons to doubt "science." As we have seen, the human experience of aging tends to lead reasonable, thinking, and intelligent people to scientifically incorrect conclusions.

The Fountain of Youth

Ponce de Leon (1460 – 1521) was a Spanish explorer commissioned by King Ferdinand II of Spain to explore North America and search for the legendary Fountain of Youth, said to restore youth and vigor to any who drank there. He did conquer Puerto Rico and named Florida around 1513, but, needless to say, never found a fountain of youth thus depriving Spain of what would have been the most profitable bottled water franchise ever. Ponce was wounded by angry Native Americans and died in Havana in 1521. School children are now taught about this amazing major exploration project that was funded by such a foolish belief on the part of the king of Spain.

Figure 25 Ponce de Leon

A "search for a fountain of youth" has ever after come to symbolize a fundamentally foolish undertaking, (especially a foolish governmental undertaking) similar to but more profound than a "wild goose chase."

Traditional theorists, especially followers of Williams' antagonistic pleiotropy theory, often refer to the fountain of youth in connection with any theory or effort directed toward serious anti-aging research. You will recall from Chapter 4 that Williams himself referred to the fountain of youth and thought that any serious anti-aging treatment was impossible.

Medical research is a "zero sum game." Any funds that are allocated for any particular research subject are presumed to be subtracted from funds available for other subjects.

Good and Evil

Most people tend to think of life, evolution, construction, and order as "good" and to think of destruction, decay, disorder, entropy, disease, aging, and death as "evil." (Entropy is often referred to as a "devil" or "demon.") We therefore naturally tend to think of animal parameters such as puberty age, gestation period, fertile intervals, and other aspects of animals obviously associated with birth and life as evolved characteristics. We also naturally tend to associate aging and menopause with "disease" and other "non-evolved" characteristics. These natural tendencies line up nicely with traditional theories of aging and no doubt had a significant role in encouraging people to look for non-evolved explanations for aging.

Aging Attitudes Survey

Aging is a phenomenon that is going to affect almost everybody. Almost every adult, (certainly almost everybody over the age of 35), therefore has an opinion or attitude regarding aging. Popular attitudes in a free-market democracy in turn affect educators, legislators, research appropriations, and career choices.

In order to get a feel for popular attitudes regarding aging, the author conducted an informal "Internet survey" regarding aging attitudes and knowledge using the search service *SeekOn* in early 2003. Details and methodology of the survey can be found in Appendix 2.

The survey was conducted as a multi-page multiple-choice questionnaire. Age, sex, educational level, and degree of exposure to training in biology were requested. Questions regarding the respondent's attitudes regarding the cause and potential for treatment of aging were asked, answered, and recorded *prior* to displaying questions regarding the respondent's knowledge about discoveries which might reasonably influence a person's attitude such as caloric restriction effects, non-aging animals, and aging genes, so that the effect on attitude from knowledge of these things could be assessed.

In addition to age, sex, and educational level we asked about training in biology because college level biology training includes information on the traditional theories of aging.

Survey Question [percent giving indicated answer]:

Have you ever studied biology? [19%]No. [58%]Yes, High School only. [23%]College

This information was used to determine how answers to questions about aging varied with degree of exposure to biology education. Since high school biology generally does not discuss aging in any detail and because most people have been exposed to high school biology, we combined "No" and "Yes, high school only" in the following breakdowns of popular attitudes.

Popular Attitudes about Aging

Here are the results of the survey:

Question: What do you think is the most likely cause of aging?
Answers (percent) vs. amount of biology education:

Answer	No College Biology	College Biology	All
All living things eventually wear out.	27	24	28
Damage to cells, DNA, or other critical function gradually accumulates.	27	36	29
We are designed to age.	41	32	36
Nobody knows. We may never know.	4	8	7

Although nearly everybody thinks the cause of aging is known, responses as to the actual cause of aging were greatly split between the three causes offered. About two thirds of the public believes in either the "wear out" theory or the "we are designed to age" theory both of which are not currently scientifically popular. People who took college level biology are more likely to believe the traditional "damage" explanations and disbelieve the "wear out" theory although 36 percent surprisingly believed the adaptive "we are designed to age" theory.

In actuality, as described in this book, we really do not know for certain what causes aging and there is substantial disagreement regarding various unproved scientific theories of aging.

Question: Which of the following most describes your views about anti-aging treatments? Answers (percent) vs. biology education:

Answer	No College Biology	College Biology	All
1. Aging is an inescapable biological reality – There will never be meaningful treatment of the fundamental causes.	61	64	62
2. Some day in the very distant future they might find a treatment.	19	12	18
3. Treatment of the fundamental causes is possible in the relatively near term.	4	12	6
4. A major treatment of aging might be as easy to do as a major treatment for AIDS.	7	8	7
5. Effective, significant treatments are already available such as HGH.	9	4	7

This question essentially asks the respondent about his or her optimism (on a scale of 1 to 5) regarding meaningful treatment of aging.

A very large proportion (80 percent) of the public believes that meaningful anti-aging treatment is either impossible or only a very long-term possibility. Those with college-level biology training were nearly as pessimistic with 76 percent of that group believing that meaningful anti-aging treatment is either impossible or very distant.

Of course this is a self-fulfilling prophecy. Those that think anti-aging treatments are impossible will not look for such treatments, will not support funding such research, and will not consider a career in anti-aging research.

The author believes the discoveries and theoretical work described in this book support a position of approximately 3.5 on the optimism scale.

Public Knowledge About Aging

The survey asked several questions regarding respondent's knowledge of various discoveries that suggest that aging is not universal or that aging can be contravened under some conditions

-Did you know that there are species that apparently do not age such as yellow-eye rockfish and some turtles? [26%] Yes [74%] No

-Did you know that genes have been found in mice and other organisms that apparently cause aging. Inactivation of these genes through genetic engineering has extended average lifespans by as much as 50 percent. [30%] Yes [70%] No

-Did you know that restricting caloric intake of lab rats while maintaining a nutritious diet has extended average lifespans by as much as 50 percent? The rats are healthier in addition to living longer. Similar results have been observed in other animals. [52%] Yes [48%] No

-Did you know that researchers are searching for a medication that would mime the anti-aging effects of caloric restriction without having to actually restrict consumption? Preliminary results are encouraging. [22%] Yes [78%] No

-Did you know that the diseases causing the largest numbers of fatalities are all age related? Ninety percent of Americans who died in 1999 were over 57. [48%] Yes [52%] No

Anti-Aging Morality and Ethics

Some of the people surveyed had moral issues with anti-aging research. When asked: "Do you think anti-aging research has any moral issues?" answers were as follows:

[43%] No

[35%] I am somewhat concerned

[22%] Yes, we should not try to extend natural lifespan.

Medicine and health care are replete with moral and ethical issues and anti-aging research or treatment is no exception.

Many people are concerned that medical advances could result in significantly adding to the fraction of a person's life spent in the "nursing home" stage. Most people would not see such a result from anti-aging research as helpful. This is a legitimate ethical concern but applies equally well to many medical activities.

Is it really moral to try to extend normal lifespan when so many people don't have an opportunity for a "normal" life because of diseases and conditions that still defy totally successful treatment?

Would seeking anti-aging treatments be seen as "playing God" to a greater extent than many other medical issues are seen as "playing God"?

A charitable organization for "anti-aging research" might fare poorly against similar organizations for "heart disease research" or "cancer research" at least partly because of moral and ethical issues.

An emerging medical ethics issue is the degree to which medical intervention should be used to alter more or less "normal" conditions. If your child is "pathologically" short, say shorter than 99 percent of the population, a physician will have no ethical problems in providing treatment. If you merely want to have a basketball player in the family, most physicians will have a problem. Is treatment of aging treatment of a "normal" condition? On the other hand, medical advances have extended average lifespan in developed countries over 100 percent in the last 150 years. Few would consider that undesirable.

Very few people have any moral or ethical issues regarding research on treatments for heart disease or cancer even though these diseases strike primarily older people and are in effect manifestations of aging. Anti-aging research could well result in better understanding of these and other age-related conditions that would help relatively younger victims. (More on this connection under Budget.)

However, notice that nearly half of respondents had no moral issues and 78 percent did not have serious moral issues with anti-aging research. In the author's opinion, moral and ethical issues are not as significant regarding public attitude toward anti-aging research as the "feasibility" aspects.

Public Opinions on Anti-Aging Research

We asked respondents about anti-aging research as follows:

> The National Institute of Aging (NIA)(part of the U.S. National Institutes of Health (NIH)) provides funds to study fundamental causes of aging as well as study of some specific age related diseases such as Alzheimer's. In 2003 NIA's budget request was about $965 million. The study of AIDS was funded at $2.8 billion. Total NIH budget was about $27 billion. For comparison, expenditures for chewing gum in the U.S. are about $2 billion annually.
>
> -Do you think taxpayer provided funding for fundamental research on aging should be: [32%] Increased [26%] Decreased [42%] Stay the Same

An obvious question is to ask how this response varied with the age of the respondent, as young people probably don't care as much about aging as older people.

Goldsmith

Age of Respondent	Increased	Reduced	Same
Under 20	25	50	25
21 - 30	16	32	52
31 - 45	15	15	70
46 - 55	36	29	35
56 - 65	62	12	26
Over 65	35	31	33
All	32	26	42

As expected, older people are more in favor of anti-aging research.

Another issue is the correlation of knowledge about various discoveries with attitude regarding anti-aging research. Of the people who had any knowledge of *any of* the indicated aspects of aging, the portion that wanted funding increased *exceeded* the average for all respondents.

Answered Yes to knowledge of:	Increased	Reduced	Same
Most deaths are aging related	46	17	37
Mimetic Research	65	18	18
Caloric Restriction	41	18	41
Aging Genes	45	9	46
Non-Aging Species	40	15	45
All	32	26	42

10. Anti-Aging Research

We have spent a lot of time discussing rather theoretical considerations such as the fundamental nature of aging. Now we can turn to more practical matters such as question 2 from the introduction:

> *Are there potentially treatable factors that are common to many different manifestations of aging?*

This is indeed the "64 billion dollar question." Medicine is currently largely occupied with finding treatments for the various individual manifestations of aging ranging from heart disease (Lipitor) to skin conditions (Botox). If there are treatable common factors, an entirely different approach could be added to current medical practice. We can define *anti-aging medicine* as treatments capable of simultaneously reducing the severity of two or more otherwise apparently unrelated manifestations of aging, and *anti-aging research* as research directed at finding such treatments or agents.

We can list the following possible answers to the commonality question:

-Yes, there are likely to be treatable common factors.

-No, any common factors are fundamental, unalterable, and therefore untreatable properties of life.

-No, there are no common factors. Individual manifestations of aging are entirely the result of many separate, independent processes that require separate treatment.

-No, any common factors are associated with necessary functions and therefore cannot be altered without unacceptable side effects.

This chapter discusses these questions and other practical considerations regarding anti-aging research.

Evolution Issues

If we consider any *one* of the apparent discrepancies between observed organism characteristics and orthodox Darwinism (e.g. aging) we find the following situation: Theories will have been produced ("orthodox explanations") that explain the discrepancy in the context of traditional evolutionary mechanics theory. Alternate theories depending on post-Darwin alternate evolution theories also exist. So, on one side we have an orthodox theory that might be convoluted, tortuous, and implausible, have logical flaws and difficulties with experimental confirmation. On the other side we might have a theory that is more plausible but depends on an alternate evolution concept that itself has credibility issues. A jury of impartial scientists might well decide that the preponderance of evidence favored the orthodox theory, and that eventually the flaws might be resolved.

If on the other hand, we consider *all* the discrepancies, then the case for alternate evolution and associated theories is dramatically improved. This book is an attempt to provide a summary of that situation.

Here is an example: There is a popular orthodox theory[39] that says that sexual reproduction and all the other evolved characteristics that produce and maintain variation, exist because they improve resistance to pathogens, an individual advantage that is said to offset all the individual disadvantages. Variation produces organisms that have slightly different immunity characteristics. As you read earlier, an evolvability theory says that sexual reproduction and other variation characteristics exist because they essentially enable the whole process of evolution in complex organisms.

As you read in chapter 5, immunity is largely independent of variation. A population of genetically identical clones would still be able to develop individual immunities. The orthodox explanation is implausible. Believing that sexual reproduction, etc. exist for the sole purpose of improving immunity is like believing that the engine in a Ferrari exists for the sole purpose of providing power to the windshield wipers. However, if you believe that alternate evolution is impossible, then an implausible explanation is better than nothing.

Evolutionary Biology

Genetics is "hot science." Companies are being formed. IPOs are being held. Billions are being made (or lost). People recognize that genetics is going to lead to major new developments in the health field as well as in development of new transgenic and genetically engineered plants and animals. The patent rights for a single important process called polymerase chain reaction (PCR) are said to have sold for $300 million.

Genetics is also relatively "hard" science as biology goes. Experiments are repeatable. Progress has been rapid. Opportunities are many. Many of the "best and brightest" go into genetics.

In stark contrast, evolutionary biology is "academic" science. Relatively few people care. There is little money in evolutionary biology.

At the same time, evolution is extremely difficult science. The Earth is thought to have existed for about 4.5 billion years. Life on Earth has been evolving for nearly 4 billion years. Of this period, good scientific records have been kept for about 200 years. Photography has only been available for about 150 years. We only have good, direct, recorded observations for about .000005 percent of the process! Fossils convey only a tiny fraction of the information that could be extracted from observations of living organisms. Although there are detailed observations of recent humans, humans are not "wild" animals and are therefore not as subject to natural selection as wild animals. Detailed observations of wild animals are difficult and expensive to do without disturbing the "wildness" of the animals. Data is sparse. Funding is limited. Progress has been glacial. *Origin* is still the most respected work. Darwin is still the most respected researcher.

Some of the theoretical modifications to Darwin's theory have very limited scope and impact. Specifically, group selection theory and selfish gene theory have been mainly used to explain a few obscure behaviors of wild animals. Few are interested in wild animal behaviors, a "hyper-academic" subject. Wild animal behaviors are subject to interpretation: Is that gorilla displaying aggression or merely scratching an itch? Many people have difficulty attaching much significance to these proposed adjustments.

The relative lack of interest (read funding) and extreme difficulty lead to a situation where evolutionary biology has aspects that appear to outsiders to resemble those of religion or philosophy. There are groups who "believe" in certain theories; other groups believe other theories. Successive generations of theorists labor to adapt their theories to new discoveries without violating their basic creeds.

Darwin picked the "low fruit." His theory of natural selection has wide, nearly universal, scientific support. Natural selection not only explains many things but also has very simple, easily understood, mechanics: "survival of the fittest."

Advancing beyond Darwin will require dealing with finer, subtler, less apparent, and less accessible details of the evolutionary process that are inevitably going to be more complex and harder to demonstrate.

As we have seen in the rat's tail exercise, proving even Darwin's theories with regard to specific cases can be difficult. The difficulty of proving more detailed theories including theories of aging, group selection, and theories of behavior (mating rituals, altruism, etc.), has hindered researchers for 150 years.

As an example of theory dealing with finer detail, evolvability deals with the capacity of organisms to evolve and therefore their future as opposed to their present. The evolvability theory of aging predicts that non-aging species such as sturgeon have an evolutionary disadvantage. If we had a time machine and could determine what descendants non-aging sturgeon and comparable aging fish had produced several hundred thousand years from now we could prove the theory. Otherwise, proving it will be difficult.

The fact that Darwin's theory has endured for 150 years tends to give it some aspects of a religion. Witness all the statements to the effect that this or that is "impossible" because it conflicts with Darwin's mechanics.

It is therefore apparent that solutions to the question "what causes aging" may be as unlikely to come from evolutionary biology alone in the next 150 years, as they have been in the previous 150 years. Even some traditional biologists such as L. Gavrilov[40] caution against basing research decisions on theories to an excessive extent:

> "Now when the single-gene life-extending mutations have been found, evolutionary biologists are presented with the task of reconciling these new discoveries with the [traditional] evolutionary theory of aging and no doubt they will ultimately succeed. However, gerontologists will also have to learn a lesson from the damage caused by decades of misguided research, when the search for major life-extending mutations and other life-extension interventions was equated by evolutionary biologists to a construction of perpetual motion machines."

Future medical researchers tend to be unaware of this history when learning in "Biology 101" about theories of aging.

The Indicator Problem

If a researcher is looking for an agent to treat an infectious disease, he can try various agents on the infectious organism in a test tube or Petri dish and within hours or days determine how effective they are in killing the infectious organism. Eventually, animal trials (assuming there is an animal susceptible to the same or similar infection) can be done in which relatively short tests of blood or tissue are good *indicators* of the effectiveness of the treatment. (Genetically engineered animals have been designed to be susceptible to human diseases they would not ordinarily develop for just this purpose.)

Unfortunately, in the case of aging, there is no such generally accepted indicator that reliably indicates the progress of aging. There is no blood test or tissue examination that can rapidly determine whether any anti-aging agent is or is not effective. A researcher having an anti-aging agent or protocol that he wanted to try could give it to a statistically significant number of rats, wait for them to die, and then determine if there is a statistically significant effect on average or maximum lifespan. This is obviously a slow process.

If rat trials were successful, the researcher could try monkeys. However, monkeys are larger, more expensive and live longer. A single trial could be expensive and take more than 20 years to complete. Finally, human trials, which might take decades to perform, could be conducted.

Clearly a reliable indicator of "aging" needs to be discovered to support anti-aging research.

This is a somewhat circular situation. If, for example, we knew that changes in the concentration of a certain hormone, or hormones were a reliable indication of aging we would have a major clue as to the operation of the aging mechanism that would probably itself lead to a treatment.

Aging Damage Mechanisms

There have been several biological mechanisms implicated in aging.

Telomeres (see Genetics) are seen to shorten on successive duplications of chromosomes during cell division. Eventually, the telomeres are sufficiently degraded that cell division via mitosis is inhibited. Shortening of telomeres was at one time seen as a "cause" of aging.

However, various types of cells in humans normally divide at differing rates. Blood cells are replaced frequently. Brain cells last essentially for the life of the person. Division of cells is obviously biologically very tightly controlled in the growth and subsequent life of an animal. Cancer is the result of uncontrolled cell growth.

It was eventually found that some cells, under some conditions, used an enzyme called *telomerase* to *repair telomeres* and therefore allow further cell division. Therefore, the absence of telomerase might cause aging. What causes telomerase to be absent?

Reactive Oxygen Species (ROS) are forms of oxygen containing compounds that are potentially destructive to tissue. ROS could obviously be part of an aging mechanism. Many people take vitamin E or other "anti-oxidant" in an effort to ameliorate the oxidation effect.

However, it was eventually found that some cells in some conditions have mechanisms to repair oxidation damage or to prevent more dangerous oxidants from forming. What determines if a cell will have repair or prevention capability?

Heat Shock Proteins are proteins that are used by cells to repair other proteins that have been damaged. Heat shock proteins are in turn produced by presence and activity of a specific protein called heat shock factor- I. Activity of HSF-I declines with age.

Causes and Effectors

Any or all of the mechanisms listed in the previous section may well be involved in causing aging. Telomere shortening, reactive oxygen species, or decline in heat shock proteins may indeed be the *proximal causes* or *effectors* of the degradation that causes aging, that is, the most immediate precursor to the deleterious effect. However, in all three cases we already know that there must be *at least* one more step in the *control of* the effector. Telomerase *controls* whether or not telomeres will be shortened. Some cells can repair oxidation damage, and so on. If it were not for the additional step, these mechanisms would affect cells more equally.

The question is: how complex is the control mechanism? Is it something like A controls B which controls C which would certainly be as complex as we would expect from a mechanism which resulted by accident (i.e. traditional theory). Maybe something controls telomerase, which control telomeres, which control aging.

If, on the other hand aging is an *evolved* adaptive mechanism (i.e. Weismann and other adaptive theories) then control of the aging mechanism is likely to be similar to *other* evolved biological mechanisms, which tend to be extremely complex. The control mechanism would likely be something at least as complicated as A controls B which controls C which controls D which controls E which controls F which controls G which finally controls telomerase which controls aging.

The control mechanism could be capable of *logic* in which multiple *inputs* are processed, something like (A OR B) AND (C OR D) controls E which controls F which controls telomerase, etc. Animals have many such complex control mechanisms as described below.

The existence of such a complex control mechanism in aging would explain interactions such as a relationship between food intake and aging or a relationship between reproduction and aging.

Biological Control Systems - Hormones

More complex animals have extensive biological control systems in the form of the endocrine system and hormones. Hormones are chemical signals that are produced by glands and other tissues that then (mostly) circulate in the blood and control biological functions in other cells. These functions in some cases involve the production of yet more hormones that then form parts of a more complex logic structure. At least 50 human hormones have been identified and more are expected. It is thought that all cells are affected by at least some hormones but usually particular hormones affect specific *target* cells.

The endocrine system is connected to the nervous system. For example, fright, nervousness, or merely embarrassment signal production of hormones that cause increases in heart rate, blood vessel changes, changes in digestive system, and so on. Detection of light is thought to affect hormones that vary on a daily basis. Some hormones vary on a monthly cycle. Some vary on a life-long cycle.

As everybody knows from their own fright responses, hormones can respond very rapidly (within seconds) to signals from the nervous system. Many hormones are seen to appear in pulses of greatly increased concentration separated by periods of relatively less concentration. The size and frequency of the pulses as well as average concentration of a hormone are all presumably significant.

Hormones are very heavily involved in growth, the reproduction process, food acquisition (hunger), digestion, nutrient utilization, and many other normal biological processes.

Although glands such as ovaries, testes, thyroid, parathyroid, hypothalamus, pancreas, pituitary, and adrenal gland, produce many hormones, some are produced in other tissue such as stomach, heart, kidney, liver and even skin and fat cells.

This system is capable of extensive logic as described in the previous section. *Negative feedback* in such biological logic "circuits" is common. In negative feedback, a hormone produced near the effector end of a control sequence affects the cells that produced a different hormone near the start of a signaling chain thus inhibiting further production of the initial hormone. Many hormones work in such pairs or even larger groups of different hormones. Hormones produced by different glands and tissues are known to interact in complex ways.

Negative feedback is almost always a part of mechanical or electronic control systems. A household furnace thermostat is an example of a control system that incorporates negative feedback.

Notice that hormone signals can either enhance or inhibit a given function. In the context of aging, a hormone might cause a tissue to age. However, a hormone could also inhibit aging in a tissue that was otherwise programmed to age. This scenario fits with observations such as progeria.

Observed concentrations of many hormones vary with age.

Many hormones (such as insulin, growth hormone, growth hormone releasing hormone, insulin-like growth factor, and prolactin) are proteins or peptides, which, since they are destroyed by the digestive system, cannot be taken by mouth. Other hormones (such as testosterone, estrogen, and progesterone) are steroids and can be taken by mouth.

Research (See Aging Genes) has indicated that a complex mechanism involving hormones controls at least some aging in the roundworm.

In their paper, *The endocrine regulation of aging by insulin-like signals[41]*, Bartke and Antebi of the Department of Ecology and Evolutionary Biology at Brown University say:

> "Reduced signaling of insulin-like peptides increases the life-span of nematodes, flies, and rodents. In the nematode and the fly, secondary hormones downstream of insulin-like signaling appear to regulate aging. In mammals, the order in which the hormones act is unresolved because insulin, insulin-like growth factor-1, growth hormone, and thyroid hormones are interdependent. In all species examined to date, endocrine manipulations can slow aging without concurrent costs in reproduction, but with inevitable increases in stress resistance. Despite the similarities among mammals and invertebrates in insulin-like peptides and their signal cascade, more research is needed to determine whether these signals control aging in the same way in all the species by the same mechanism."

Cynthia Kenyon is a researcher at the University of California in San Francisco and has performed research on the roundworm that indicates that its aging is controlled by a complex mechanism:

"Studies from our lab have led to the discovery that aging in *C. elegans* is controlled hormonally by an insulin-IgF/1-like signaling system. Mutations in genes that encode components of this system double the lifespan of the animal and keep it active and youthful much longer than normal. This system is regulated by environmental cues and signals from the reproductive system. We have identified many new long-lived mutants and hope to identify new genes and steps in the aging pathway."

Involvement of hormones is a major clue that aging is an evolved trait controlled by a complex mechanism. See more under Hormones and Aging Genes.

Biological Clocks

What controls the sequencing of longer-term biological functions? If aging is part of an organism design feature that controls lifespan, what is the ultimate source of the timing function that controls life-cycle functions such as puberty, menopause, and aging? Some theorists think that long-term life-cycle functions are timed by an internal cell clock based on some very gradual effect such as telomere shortening. According to this concept, all the cells are individually running "biological clocks." However, the central theme in any multi-cell organism is *coordination*. All those cells are somehow assembled to produce specific structures and then execute particular functions that are coordinated by signaling, sometimes organism-wide signaling by means of hormones or nerve signals. If aging is a biological function there is no reason to believe that it would be coordinated in a different manner from other life-cycle functions such as puberty. Few would argue against the idea that puberty and other reproductive functions are coordinated by hormones.

So what is the ultimate source of the "clock?" Is there some central cell "counting" function (perhaps in some specific gland or tissue) that controls hormones that then control life-cycle functions? Many organism life-cycle functions are clearly synchronized to and even "phase-locked" to external event cycles. For example, many organisms (including mammals) have reproductive cycles that not only occur on an annual basis but occur at a particular time during the year. Obviously the organism has some way of detecting the external cycle and synchronizing the reproductive functions. Could *other* life-cycle functions be based on similar external cues? Could lifespan, puberty, and other life-cycle functions be derived by counting external cycles?

This concept suggests an animal experiment. Rats, mice, or other short-lived organism could be maintained under conditions that simulate longer and shorter daily and annual cycles. We could have day-night cycles of perhaps 19 and 29 hours and determine if animals kept under these conditions had longer or shorter lifespans, earlier or later ages at puberty, or other differences in their development cycles.

Impact of Theories

Theories of aging drastically influence anti-aging research in two different ways.

First, the "optimism" of a theory obviously influences decisions regarding investment of research resources by its followers. Few legislators and administrators want to invest in foolish research. Few researchers want to embark on a career in which significant advances are widely thought to be impossible or extremely unlikely.

Second, the *directions* in which research is conducted are highly influenced by the theories respected by the researchers. Many age-related research proposals begin with something like: "According to the [whatever] theory of aging…" Followers of traditional theories are not likely to consider complex, obviously evolved, aging mechanisms.

Regarding research direction, the most important choice resulting from Darwin's dilemma seems to be "adaptive" or "non-adaptive" (or if you prefer, programmed or non-programmed; active or passive) The actual basis or genesis of a theory seems to be less important in determining the direction taken by researchers. If aging is "adaptive" then researchers are going to be looking at complex mechanisms that offer the capability for mediating the observed interrelationships between aging and food supply, aging and reproduction, etc. If aging is "non-adaptive" then researchers are going to be looking at actual effectors such as telomere shortening, or processes immediately prior to effectors that could plausibly result from accidental mutations or side effects.

Regarding "optimism", the non-adaptive theories, for reasons enumerated in Chapter 4, are generally pessimistic with regard to the possibility of major medical intervention in aging. The antagonistic pleiotropy theory, in particular, specifically teaches that major medical intervention in aging is impossible. If aging is non-adaptive, then it is a problem that 4 billion years of evolution have been unable to fix, a very, very difficult problem indeed!

If aging is an evolved adaptation, then prospects for successful medical intervention are dramatically better for a number of reasons.

Goldsmith

If aging is an evolved trait, and similar to other similar evolved traits, it almost certainly involves a very complex control mechanism. Medical intervention in aging would require developing a way to interfere with *any one* of the many parts of the aging mechanism *without* interfering significantly in the operations of the myriad other biological mechanisms we need to live happily. This is a familiar problem in medicine. Chemotherapy involves finding agents that have the maximum adverse effect on cancer cells with the minimum adverse effect on healthy cells. Fighting infectious diseases involves finding agents or procedures that interfere with the life processes of the infecting organism with minimum effect on the host. The more complex and centrally controlled the aging mechanism is, the more likely it is that such an attack point or points can be found.

Another optimistic aspect is that complex control mechanisms in animals usually involve hormones. Although there are at least fifty human hormones, most hormones are "exposed" and accessible in that they are found in the blood, can be measured with relatively little invasion, and can be synthesized, injected or (sometimes) orally administered, and so forth.

Anti-Aging Quacks and Scams

Aging, as a universal affliction, is an obvious favorite of quacks, charlatans, and scam artists and has been for hundreds and probably thousands of years. This is no doubt part of the reason for the deep and, so far, well-deserved, skepticism most people have regarding the possibility of significant anti-aging treatments (and associated research). Because of the progress medical science has made concerning other afflictions, we can expect that aging will become an increasing target for quacks and scammers.

Anyone who uses the Internet is familiar with the countless "spam" e-mail messages touting this or that anti-aging remedy:

Says one: "Scientific Breakthrough", "Human Growth Hormone Therapy", "Lose weight while building lean muscle mass and reversing the ravages of aging all at once."

These products will (allegedly) do more than even the snake oil or monkey glands of yore: "Lose Weight, Build Muscle Tone, Reverse Aging, Increased Libido and Duration Of Penile Erection, New Hair Growth, Improved Memory, Improved skin, Wrinkle Disappearance, and more." Many claim to do it all "while you sleep", with "no exercise required."

Most of the heavily advertised remedies involve human growth hormone (HGH).

HGH is a protein and therefore cannot be taken by mouth. Some of the remedies therefore claim to stimulate the body to produce more HGH.

HGH is one of the hormones that decline with age. Some clinical trials have been conducted regarding use of HGH in elderly patients without notable success. Although HGH has been seen to increase lean muscle mass and bone density, actual beneficial effect such as increased muscle strength has not been demonstrated.

There is no scientific evidence that HGH is beneficial in connection with the large number of claimed benefits. Animal trials have not indicated success in increasing maximum or average lifespan.

A steroid hormone, DHEA (dehydroepiandrosterone), a precursor to some other steroid hormones such as testosterone, which also decreases with age, has received interest as an anti-aging agent. Some positive effects have been observed in elderly patients but wild claims are greatly overblown.

Hormones, according to adaptive theory, are almost certainly involved in human aging, and have been demonstrated to be involved in the aging of some organisms. Hormone therapy of some sort may eventually be a significant treatment for aging. However, aging is almost certain to involve more than one hormone, possibly many hormones, possibly including a currently undiscovered hormone.

Nothing in this book should be interpreted as endorsement of any currently available anti-aging medication with the possible exception of statins and aspirin.

Physicians and other health professionals have a unique situation regarding aging.

The average person has to deal "up close and personal" with the severe effects of aging only when they or a close relative reach an advanced age. Most physicians have to deal with aging on a daily basis. The practitioners of any profession need to learn to accept the limitations of their profession and aging is arguably the greatest single constraint on the practice of modern medicine.

All physicians took Biology 101 and were likely exposed to training to the effect that significant medical intervention in aging is impossible.

Physicians are intimately familiar with the human experience of aging and generally far less familiar with other mammals, rockfish, salmon, and bamboo, and their implications for aging theory. The human experience suggests that significant intervention in the aging process itself is impossible and that we are limited to treating individual manifestations.

Frauds are not confined to relatively low-end efforts on the Internet. There are licensed physicians selling all sorts of purported anti-aging treatments having little or no clinically demonstrated effectiveness. To be fair, clinical demonstration of the effectiveness of an anti-aging treatment is unusually difficult for reasons mentioned in Chapter 1.

For these reasons many, possibly most, physicians understandably consider "anti-aging medicine" to be equivalent to "quackery." If you show this book to your family doctor, you may well get such a reaction. This has a major and obvious negative effect on funding and pursuit of anti-aging research. Medical research is largely and appropriately controlled by medical people.

Strategies for Engineered Negligible Senescence (SENS)

Aubrey de Grey, based in Cambridge, England, is a colorful figure in anti-aging research and a founder of the SENS Foundation. *Strategies for Engineered Negligible Senescence (SENS)* is founded on the idea that "Human Regenerative Engineering" is possible and that human aging can not only be stopped but can be reversed:

"A key aspect of SENS is its potential to extend healthy lifespan without limit, even with repair processes which remain imperfect, as the repair only needs to approach perfection rapidly enough to keep the overall level of damage below pathogenic levels."

SENS is a descendent of the earlier but still existing *Methuselah Foundation*. De Grey is Chief Science Officer of SENS and also chief editor of a journal *Rejuvenation Research* described as "the world's only peer-reviewed journal focused on intervention in aging" (no longer true – see *Phenoptosis* below). SENS holds periodic symposia.

De Grey is certainly seen as part of the radical fringe by many main-line gerontologists for his view that immortality is possible. However, his journal has a respectable impact factor compared to other gerontology journals and attracts articles from respected researchers.

While holding many radical views regarding aging, de Grey believes in the traditional passive accumulation of damage theories that are essential to his SENS damage-oriented view of the aging process. The SENS approach "breaks the aging problem down into seven major classes of damage and identifies detailed approaches to addressing each one." More specifically, de Grey believes aging is the result of inadequacies in a potentially large number of biological maintenance and repair mechanisms. As described in the next section, which compares active and passive maintenance theories, active (programmed aging) theories provide a much better match to observations.

Programmed and Non-Programmed Maintenance Theories of Aging Compared

Aubrey de Grey has advanced an argument (to my knowledge the only argument) against programmed aging that does not depend on assuming the utter invalidity of all of the post-1962 non-individual-benefit evolutionary mechanics theories and/or all of the dependent programmed aging theories. Consequently it is worth taking the time to describe the many flaws in this argument.

Active (programmed) and passive (non-programmed) maintenance and repair theories of aging both assume generic deteriorative processes (wear and tear, oxidation, telomere shortening, etc.) exist that would cause deterioration in any organism unless actively opposed by the organism. These processes include disease-specific processes such as those that result in cancer (unopposed adverse mutations), or heart disease (arterial deposits or other artery damage) that are very different and presumably involve very different maintenance and repair activities. The passive non-programmed aging theory proposes that the very large differences in lifespan between different mammals result entirely from differences in the operation of their maintenance and repair activities that act to oppose the deteriorative processes. These differences are said to result in the observed huge lifespan differences of more than 100 to 1 in mammals and more than 600 to 1 in fish.

De Grey's argument is that active and passive mechanisms both result in organisms having the specific lifespan needed by a particular species and that they are consequently functionally identical. If this were true (it is not), there would be no evolutionary motivation to evolve and retain the more complex lifespan regulation mechanisms proposed by proponents of programmed aging. De Grey proposes that his aging *mechanism* concept (which is compatible with non-programmed aging theories) is valid even if programmed aging theories and their underlying evolutionary mechanics concepts such as group selection or evolvability are valid. De Grey's argument is directed *only* at the observed inter-species lifespan differences and ignores *many* other observations. It also has many logical issues described below. De Grey also ostentatiously ignores requirements and predictions of specific programmed aging theories in his claim that his passive mechanism would be functionally identical to an active mechanism in satisfying a programmed aging theory.

In the passive maintenance and repair concept suggested by de Grey and shown below, aging is ultimately the result of generic deteriorative processes such as oxidation, molecular disruption, genetic transcription faults, mechanical damage, and other natural processes that cause deterioration in biological systems. The gross lifespan differences are explained by the presence of a large number of independent anti-deterioration functions that act to prevent damage from or repair damage resulting from the generic deteriorative processes. A particular longer-lived mammal species possesses more effective anti-deterioration functions than a shorter-lived species and consequently is able to slow the accumulation of damage.

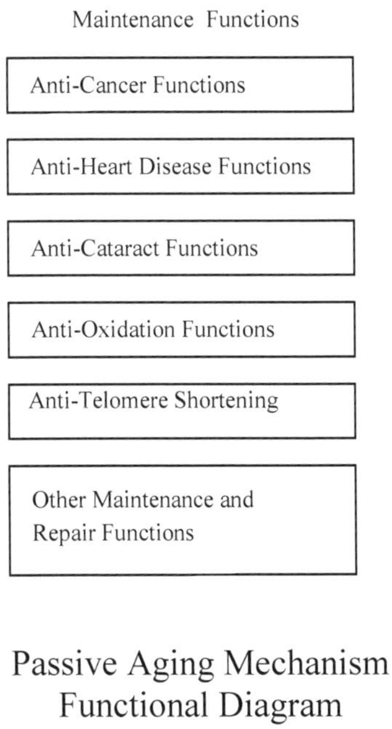

Figure 26 Passive Aging Mechanism

This concept suggests that organisms possess a potentially large number of maintenance and repair functions that are functionally independent and evolved independently in order to produce the lifespan needed by the organism. If for example, cancer at too young an age was preventing a species from obtaining the particular lifespan needed by that species, the species would evolve better anti-cancer mechanisms. If heart disease at too young an age was a problem, the species would evolve better anti-heart disease functions. It is understood that the mechanisms ultimately responsible for cancer and heart disease and other diverse manifestations of aging are very different and consequently the associated maintenance and repair mechanisms are presumably very different. If for some reason, a species needed a shorter lifespan as required by a programmed aging theory, or did not need as long a lifespan as specified by a non-programmed theory, all of its maintenance and repair mechanisms would eventually be gradually degraded by random unopposed mutations until the target lifespan was obtained. Therefore de Grey suggests his passive maintenance concept would satisfy both programmed and non-programmed theories of aging in regard to obtaining the species-specific lifespan needed by each species.

The diagram below describes an active programmed aging concept in which maintenance and repair functions are further controlled by a biological clock mechanism. The clock directs the various maintenance and repair mechanisms to decrease their effects as a species-specific function of age in order to result in the species-specific lifespan. The clock mechanism can in turn be adjusted by sensory functions that can detect and respond to external conditions that alter the optimum lifespan for the organism.

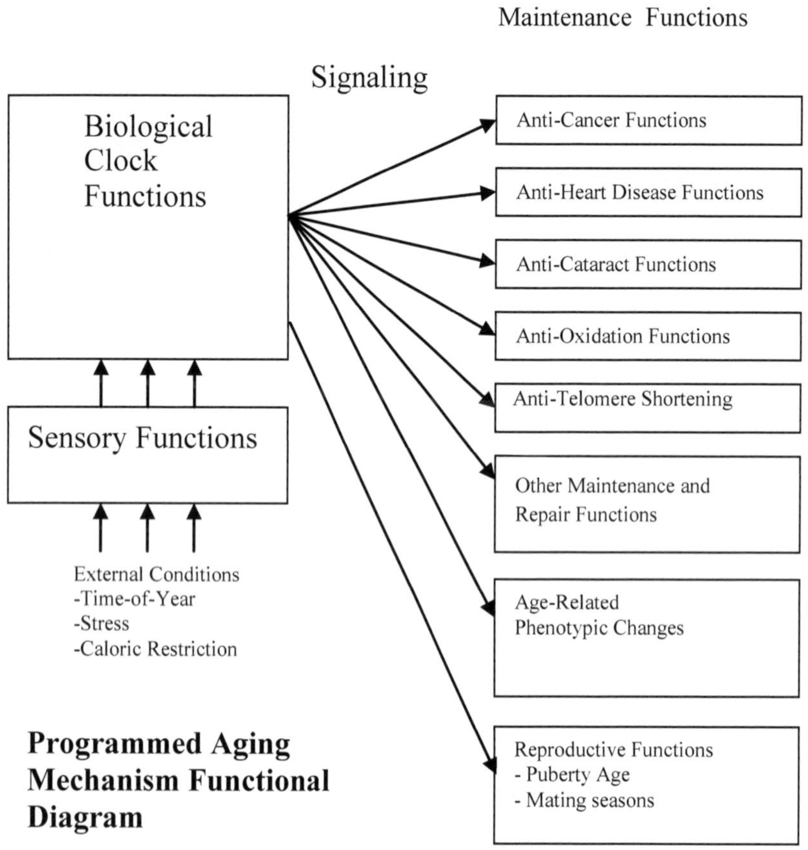

Figure 27 Programmed Active Aging Mechanism

This concept encompasses the idea that not all physiological changes associated with aging necessarily involve "damage" or generic deteriorative processes per se. An extension of the growth program could purposely program decreases in muscle strength, reductions in reproductive activity, and other phenotypic changes that are seen as adverse under traditional evolutionary mechanics because these changes create an evolutionary advantage according to a programmed theory of aging. If the program can direct increases in muscle strength and reproductive activity it could obviously also direct decreases in muscle strength and reproductive activity. Programmed decreases in some biological function, sometimes by means of programmed cell death (*apoptosis*), are common. A tadpole's tail both grows and shrinks according to its program.

As indicated on the diagram, reproductive functions such as age of initial reproductive maturity and mating seasons are also controlled by a biological clock that is adjusted by external conditions such as mating seasons that occur at a particular time of year. There is no scientific disagreement with the idea that reproductive functions are controlled by such a system. The proposal here is that an aging function that evolved because it served a purpose would logically be implemented in essentially the same manner as the reproductive functions and would also involve a biological clock and mechanisms that allow for sensing of external and internal (see below) conditions that affect optimum lifespan.

The passive aging mechanism proposed by de Grey requires a number of assumptions:

First, it assumes the existence of a potentially large number of different deteriorative processes that would cause organism deterioration unless actively opposed by the organism. If there were no deteriorative processes presumably aging would not occur according to non-programmed theories. There is no scientific disagreement with this idea and many such processes including wear and tear, oxidation, and telomere shortening have been identified.

Second, it assumes that living organisms possess a potentially large number of different maintenance and repair functions that counteract the deteriorative processes. Again there is no scientific disagreement and many such functions have been identified. Wounds heal, hair grows, and skin cells are replaced. It is generally accepted that merely maintaining life requires the expenditure of energy and resources. We need to keep breathing even when asleep.

Third, much more controversial and counter-intuitive, it assumes that the evolution process would result in each of the maintenance and repair functions *varying in effectiveness over a huge range* in order to explain the huge differences in lifespan between otherwise biochemically similar organisms. The anti-cancer mechanisms, the anti-heart disease mechanisms, the anti-cataract mechanisms, and so forth must all be different in humans than in dogs in order to explain the gross difference in the ages at which these symptoms appear.

This idea is very counter-intuitive because it would appear that maintenance and repair functions are generally binary in nature: An organism either can perform the function or it cannot. Once an organism evolved the capability for performing some maintenance and repair function, wouldn't that particular problem be solved for it and its descendant species regardless of the age of the organism? Some reptiles can replace a lost limb, mammals cannot. It is not obvious why "replace skin cells" would be a different process in an 80-year-old then in a 2-year-old or a different process in a mouse than in a man. If the process is not incrementally different why would the function need an incrementally different design as organism lifespan incrementally increases? It is easy to see how random mutations or even a single mutation could completely destroy the functioning of a complex maintenance and repair mechanism. It is much harder to see how random mutations could result in a 600 to 1 variation in the effectiveness of such a mechanism to correspond with observed variations in fish lifespan.

This difficulty is progressively more severe with shorter-term maintenance processes. If we discontinue a particular maintenance or repair process, how long does it take for the resulting organism deterioration to produce a significant degradation in survival or reproductive characteristics and therefore be significant in an evolutionary sense? We could use this as a measure of "term." Now imagine that some process has a term of one year. Perhaps in rats this process is 90 percent effective and in dogs it is 99 percent efficient, and so forth. For reasons described above this is implausible but maybe possible. The problem is that most obvious maintenance and repair functions (hair grows, wounds heal, etc.) have short terms. Sleep is widely accepted as a maintenance and repair function and has a term as defined above of about 24 hours. Are we to believe that the sleep process is 99.999 percent effective in rats and 99.99999 percent effective in humans? What would be the specific mechanics of such a process that accommodated such fine variation in response to random mutations?

If indeed there existed maintenance and repair processes that had relatively long terms we would expect differences in *symptoms* of aging between short and long lived organisms. For example, if the mechanisms that cause cancer inherently took a long time to operate, we would not expect to see cancer in relatively short-lived organisms like dogs. This has not proved to be true. Symptoms of aging are very similar between dogs and humans but occur at very different ages.

To summarize, the binary and short-term nature of maintenance and repair functions suggests that if a mammal evolved maintenance and repair mechanisms adequate for a lifespan of three years, that the identical mechanisms would be good enough for a lifespan of three hundred years! The issues described above suggest that non-programmed aging would work better in very short-lived organisms. Perhaps such an organism would have a very short lifespan because it simply did not possess relatively longer-term maintenance and repair mechanisms. This too fails to match observations: *C elegans,* with a lifespan of 21 days, has a complex programmed aging mechanism!

According to all of the evolutionary programmed and non-programmed theories of aging, the most important single factor that determines how long a particular species needs to live is the age at which it is first capable of reproducing. As described above, it is very widely accepted that this reproductive function is controlled by a biological program that in turn is capable of detecting and adjusting to external conditions such as planetary seasons. It would therefore appear to be obviously beneficial for an organism to be able to alter its lifespan to accommodate *changes in its own age of reproductive maturity* or other reproductive parameter. An aging program would be able to do this where the passive concept would not.

Many temporary or local external conditions could plausibly temporarily affect the optimum lifespan for an organism including harsh environmental conditions, predation, decreased food supply, overpopulation, and other forms of stress. An organism that could temporarily alter its lifespan in response to these external conditions would have obvious evolutionary advantages. For example, harsh environmental conditions or heavy predation could increase the externally caused mortality in a population. An organism that could increase its internally imposed lifespan to compensate would thus have an advantage. The ability of organisms to adjust biological mechanisms in response to local and temporary external conditions is very common. In connection with lifespan the programmed mechanism would have this capability while the non-programmed mechanism would not.

Observations overwhelmingly favor programmed lifespan regulation mechanisms. Genes have been discovered that cause aging. Many species such as salmon, octopus, and many semelparous species have very explicit suicide mechanisms. Examples of signaling and detection of external conditions in explicit lifespan regulation mechanisms have been discovered.

Traditionally, proponents of non-programmed mammal aging have claimed that all such instances are examples of special cases where an organism *needed* a programmed lifespan limitation according to *their* aging theory. They propose that according to *their* theory, these special cases do not apply to most mammals and that therefore humans and most mammals possess non-programmed aging mechanisms despite all the contrary evidence. They consequently dismiss examples of obviously programmed lifespan limitation in non-mammals as irrelevant to human aging.

In contrast, programmed aging theories generally contend that most complex species would benefit from a purposely limited lifespan. Some theories supply arguments that more complex organisms such as mammals have a *greater* need for a purposely limited lifespan than simpler organisms and even suggest that gradual aging would have evolutionary benefits over biological suicide or semelparity. De Grey, having desisted from claims that programmed theories and their underlying evolutionary mechanics concepts are invalid, cannot so easily dismiss these observations. If an octopus needs a suicide mechanism why would a more complex organism not need lifespan regulation? Arguing against this logic would require that de Grey argue against specific programmed aging theories or their underlying evolutionary mechanics concepts, and he has declined to do so.

The aging mechanism proposed by de Grey and other non-programmed aging proponents (multiple different maintenance and repair mechanisms that independently evolved) suggests that mechanisms behind major symptoms of aging are functionally independent and that there may be few if any *potentially treatable common factors* between otherwise unrelated symptoms of aging such as heart disease and cancer.

Programmed aging mechanisms such as described above suggest that there are many elements of the aging mechanism (clock, signals, sensing features, etc. i.e. the "program") that are common to many symptoms of aging and represent targets for intervention. Evidence overwhelmingly supports the programmed mechanism. Stress in the form of caloric restriction, or exercise[42] generally delays aging. Single-gene human genetic diseases such as Hutchinson-Guilford progeria[43] and Werner syndrome[44] greatly accelerate multiple symptoms of aging. Some pharmaceutical agents such as statins have been shown to have a beneficial effect on both heart disease and cancer.

The existence of apparently non-aging (negligibly senescent) organisms acts to confirm the idea that, once evolved, maintenance and repair mechanisms could result in at least extremely long and possibly indefinitely long lifespans. Under the programmed mechanism concept, such observations can be explained as failures in the programming mechanism (e.g. in the clock or other common element) that caused the organism to fail to limit its lifespan. According to programmed theories the negligibly senescent species would be at an evolutionary disadvantage and therefore likely to become extinct, an idea that is compatible with the relative rarity of negligibly senescent species in comparison to the much larger number of similar senescent species. Non-programmed aging proponents implausibly contend that the negligibly senescent species for some conveniently unspecified reason needed an *extremely* long life and therefore evolved *extremely* efficient maintenance and repair mechanisms. Perhaps they are 99.999999999999999 percent efficient!

Any claims that non-programmed passive aging mechanisms satisfy the need for programmed aging should make at least some attempt to describe how passive mechanisms would satisfy the many relevant observations described above as well as some attempt to satisfy the requirements of specific programmed aging theories or explain why those requirements are invalid. Programmed aging theories say a complex programmed mechanism would evolve because of the many evolutionary advantages it would convey. If this is not the case, why not? If non-programmed mechanisms are as effective as programmed mechanisms, why, pray tell, is there so much direct evidence of programmed mechanisms? These theories have grossly different predictions regarding the best way to approach treating and preventing massively age-related diseases and so need careful consideration.

Homo Sapiens Liberatus – V. P. Skulachev

Prof. Vladimir Petrovich Skulachev is the dean of Bioengineering and Bioinformatics and director of the A. N. Belozersky Institute of Physico-Chemical Biology at Moscow State University (MSU). He is also an academician in the Russian Academy of Sciences. MSU is the largest (43,000 students, 19,000 staff) and oldest (1755) university in Russia. Skulachev is arguably the foremost and most senior proponent of purposely programmed (adaptive) aging and directs various associated research activities including *Homo Sapiens Liberatus* (HSL).

The idea behind HSL is that some evolved characteristics of humans (specifically including evolved biological mechanisms that purposely limit lifespan) no longer serve any purpose in that the genetic evolution process (as previously noted in Chapter 2) no longer operates effectively in humans. We should therefore work to overcome these obsolete mechanisms.

Figure 28 Vladimir Skulachev

One major research activity in this area is the SkQ Megaproject[45]:

"Antioxidants specifically addressed to mitochondria have been studied for their ability to decelerate aging of organisms. For this purpose, a project has been established with participation of several research groups from Belozersky Institute of Physico-Chemical Biology and some other Russian research institutes as well as two groups from the USA and Sweden, with support by the "Mitotechnology" company founded by "RAInKo" company (O. V. Deripaska and Moscow State University)."

The SkQs are a family of compounds "comprising plastoquinone (an antioxidant moiety), a penetrating cation, and decane or pentane linker."

The cited article (2007) reports some exciting results:

"Extremely low concentrations of SkQ1 and SkQR1 completely arrest the H2O2-induced apoptosis in human fibroblasts and HeLa cells (for SkQ1 C1/2=1.10(-9) M). Higher concentrations of SkQ are required to block necrosis initiated by reactive oxygen species (ROS). In mice, SkQ1 decelerates the development of three types of accelerated aging (progeria) and also of normal aging, and this effect is especially demonstrative at early stages of aging. The same pattern is shown in invertebrates (drosophila and daphnia). In mammals, the effect of SkQs on aging is accompanied by inhibition of development of such age-related diseases as osteoporosis, involution of thymus, cataract, retinopathy, etc. SkQ1 manifests a strong therapeutic action on some already developed retinopathies, in particular, congenital retinal dysplasia. With drops containing 250 nM SkQ1, vision is recovered in 50 of 66 animals who became blind because of retinopathy. SkQ1-containing drops instilled in the early stage of the disease prevent the loss of sight in rabbits with experimental uveitis and restore vision to animals that had already become blind. A favorable effect is also achieved in experimental glaucoma in rabbits. Moreover, the pretreatment of rats with 0.2 nmol SkQ1 per kg body weight significantly decreases the H2O2-induced arrhythmia of the isolated heart. SkQ1 strongly reduces the damaged area in myocardial infarction or stroke and prevents the death of animals from kidney infarction. In p53-/- mice, SkQ1 decreases the ROS level in the spleen cells and inhibits appearance of lymphomas which are the main cause of death of such animals."

More recent results including some clinical trials of SkQs against various human conditions are reported in the *Homo Sapiens Liberatus Workshop* conducted in Moscow, May 2010:

http://www.programmed-aging.org/Moscow_WS_2010/

Caloric Restriction Mimetics

One effort in anti-aging research is being applied to the development of agents that would "mime" the effect of caloric restriction on aging without requiring a person to actually restrict his or her diet. The *mimetic* would fool the aging process.

In the August 2002 issue of *Scientific American*, in an article titled *The Serious Search for an Anti-Aging Pill[13]*, Lane, Ingram, and Roth report on their work in developing such a mimetic.

Caloric restriction produces measurable differences in animals including lower body temperature, lower weight, greater sensitivity to insulin, lower fasting levels of glucose and insulin, and later onset of age related diseases including cancer, in addition to longer average lifespan and longer maximum lifespan.

In their tests, the compound 2DG (2-deoxy-D-glucose) resulted in many but not all of the same physiological changes. (The effects of 2DG on maximum and average lifespan have not yet been determined.) The 2DG compound itself is not suitable for human use because the non-toxic range (difference between the effective therapeutic dose and toxic dose) is not sufficient.

This research is obviously exciting. Determination of the mechanics whereby 2DG simulates caloric restriction could well lead to development of other, safer, mimetics usable in humans or lead to understanding of aging mechanisms that could lead to development of other anti-aging agents.

Caloric restriction experiments may result in development of reliable indicators of aging (such as hormone levels).

Reversibility of Aging

If aging causes damage that is irreversible, then a theoretically perfect anti-aging medication could halt further damage but could not reverse damage that had already occurred. On the other hand, if aging does not involve irreversible damage, a perfect anti-aging agent could reverse the effects of aging in addition to halting further deterioration.

Obvious maintenance and repair mechanisms are more of a repair rather than prevention nature; hair grows, skin cells are replaced, wounds heal.

The various functional theories of aging generally do not speak to this issue since functionally, in the absence of treatment, the two cases are identical. Reversibility or irreversibility would not affect evolution. The disposable soma theory does consider the damage caused by aging to be reversible.

Mechanisms proposed by non-programmed theories (see earlier discussion) generally predict that damage monotonically increases during the life of an organism and that the variable differentiating species lifespans is the *rate* at which damage accumulates, an idea that is compatible with irreversible damage. Programmed theories tend to suggest that programmed decrease in maintenance and repair or other programmed phenotypic changes causes aging. Damage mechanisms and their associated maintenance and repair mechanisms could operate on rather short terms relative to lifespans. This concept is more repair-oriented. Programmed phenotypic changes that are not actually damage, per se, like decreases in reproductive activity are nominally reversible. In other words, programmed theories and their associated aging mechanisms are much more likely to allow damage repair; non-programmed theories are more likely to involve irreversible damage.

Reversibility could critically affect the difficulty of experimentation. If aging is substantially reversible, experimental trials of prospective anti-aging agents and protocols could be dramatically shorter than if aging is substantially irreversible. For example, in human terms the data in Chapter 1 shows that people 93 years old have an approximately 20 percent chance of dying within a year. If aging were reversible, an even moderately successful anti-aging medication administrated to people 93 years old would presumably significantly reduce death rate during a trial of only a few years. If aging were irreversible, trials would presumably have to be much longer and start at much younger ages to determine an anti-aging effect.

If we map this same relationship onto an animal (such as a rat) having a much shorter lifespan, rather short trials of prospective anti-aging agents are possible if aging is reversible.

It should be possible to assess the reversibility of aging by using caloric restriction on animals of different ages. Some investigators have published caloric restriction results indicating that aging is indeed at least somewhat reversible.

Aging Research Budget

The arm of the U.S. Government responsible for federally funded health research is the National Institutes of Health (NIH).

Here are some excerpts from the 2004 budget request for the NIH that totals $27.9 Billion:

Function	**$(M)**
Cancer	4501
Allergies, Infectious Diseases	2928
AIDS	2869
Heart, Lung, and Blood	2793
General Medicine	1869
Diabetes, Kidney, Digestive	1789
Mental Health	1200
Child Health	1114
Aging (inc. Alzheimer's, etc.)	**989**
Resource Resources	901
Eye	636
Dental	357
Library of Medicine	309
Other	5637
Total	27892

NIH FY 2014 numbers are about 11 percent higher at $31 billion. U.S. inflation during the same period was about 23 percent so NIH spending did not keep up with inflation.

NIA (Vicky Cahan) advises that 57% of the funding in the National Institute of Aging (about 1.9 percent of the NIH budget) is for basic research with the remainder assigned to research on specific age-related diseases such as Alzheimer's disease.

In contrast, Americans spend about $2 Billion on chewing gum annually.

Imagine how these numbers would change if most people believed that there actually was a reasonably short-term possibility that a major treatment for aging was possible and that such a treatment would reduce or delay the incidence of heart disease and other age-related disease. The anti-aging budget might exceed the bubble gum budget!

Potential Anti-Aging Agents

Millions of people take statin drugs (Lipitor, Zocor, Crestor, Mevacor, Pravachol, etc.) in an effort to delay heart disease. Recent clinical evidence suggests that statins also delay certain forms of cancer. A study performed by the University of Michigan[44] indicates that statins reduced the risk of developing colorectal cancer by about 50 percent. A Johns Hopkins study showed similar improvement for advanced (metastatic) prostate cancer. A Louisiana State University and Veterans Administration study showed similar risk reductions for breast, prostate, lung, and pancreatic cancer. Further study may well indicate similar reductions in risk for other, less prevalent, forms of cancer.

Cancer and heart disease are highly unrelated except that both are symptoms of aging. Therefore, statins appear to represent an actual "anti-aging drug" with some clinical support. They may attack the fundamental aging process as opposed to (coincidentally) attacking the processes of unrelated diseases. It is therefore possible that statins or similar drugs have other anti-aging properties. This is an obviously exciting development in anti-aging medicine.

Resveratrol is a component of grape skins and red wine. It has been noticed that people who drink a lot of red wine seem to live longer than otherwise similar populations and especially have reduced incidence of heart disease. Experiments with short-lived fish[46] indicate substantial lifespan increases (30 percent) in animals fed resveratrol. Some believe that taking resveratrol in capsule form may be ineffective as it is largely destroyed by digestion and therefore take the powdered form in the hope of more absorption by oral tissues.

Heterochronic Plasma Exchange Experiments

Harold Katcher is a professor of Biology at the University of Maryland.

Programmed aging theories contend that aging is a biological function that has evolved because it serves a purpose. In mammals and many other organisms, most biological functions such as reproduction, digestion, growth, fight/ flight response, and circadian, monthly, and annual cycles, involve signaling. Chemical signals distributed in the blood (e.g. hormones) coordinate activities of different tissues in order to accomplish the function. Katcher believes that signals in the blood of old mammals could direct tissues receiving the signals to age or allow aging to occur. Alternately or in addition, signals in the blood of young mammals could direct receiving tissues to activate anti-aging processes or otherwise exhibit a young phenotype.

Plasmapheresis is a procedure that has been used since the 1960s to treat various diseases. In this procedure, blood is gradually removed from a patient and processed to separate the cells from the plasma. The cells are then returned to the patient in such a way as to maintain a safe blood supply. Donor plasma or manufactured plasma substitute can accompany the return of cells to the patient in order to replace removed plasma. Donor plasma can be obtained through plasmapheresis in which only cells are returned to the donor if quantities are restricted to one liter twice weekly. Plasmapheresis is used to treat diseases in which harmful components such as certain antibodies have accumulated in the plasma. Plasmapheresis is a mature accepted process and equipment for performing the processing is available.

Katcher thinks plasmapheresis could also be used to perform a heterochronic plasma exchange between young and old humans or other mammals. In this case, plasma taken from a young individual would be used to replace plasma removed from an old individual and thus cause reversal of the signal situation and consequent change in the direction received by tissues regarding aging. Such a procedure could be important in the treatment and prevention of age-related diseases and could also eventually lead to identifying the specific blood components responsible for enabling or inhibiting the aging process with obvious potential medical applications. Substantial experimental evidence supporting this approach already exists.

11. Conclusions and Recent Developments

The Current Case for Programmed Aging in 2014

The programmed vs. non-programmed aging question is one of the most important questions in modern science. In developed countries, age-related diseases are clearly the most important public health problem. If they pick the wrong theory, legions of health researchers are going to be looking in the wrong places.

When this book was first published in 2003, it was clear that the preponderance of evidence strongly favored programmed aging. A jury composed of scientists with no personal interest in or preconception of the answer, having reviewed the then-current empirical evidence, published arguments, and state of evolutionary mechanics theory, would find programmed aging clearly more likely than non-programmed aging. Since then, a number of developments have advanced the status of programmed aging to essentially that of "beyond a reasonable doubt" as follows:

Experimental evidence favoring programmed aging has continued to accumulate.

Arguments against programmed aging are increasingly pro forma and ideological and less scientific in nature. Specific counter-arguments against specific modern programmed aging theories or underlying modern non-individual-benefit evolutionary mechanics theories have not appeared.

Specific arguments against non-programmed aging theories continue to appear[47].

Leaders of the non-programmed aging faction, while still objecting, have made important concessions that collectively support programmed aging.

Specific proposals have been advanced regarding multiple evolutionary mechanics (propagation) concepts that support evolution and retention of programmed aging.

A new widely indexed journal dedicated to the assumption that programmed aging is the correct theory has begun publication.

Evolutionary Mechanics Summary

Aging theories are essentially dictated by the underlying evolutionary mechanics concept.

Modern "evolutionary" non-programmed theories are based on orthodox Darwinian individual-benefit-only evolutionary mechanics as modified by Medawar's population-oriented concept. Most modern theorists also accept Williams' ideas: The evolutionary disadvantage of aging must be effectively zero; aging must convey an evolutionary advantage that compensates for its individual disadvantage; therefore aging must be linked to an individually advantageous trait or traits. This evolutionary mechanics concept leads to the conclusion that programmed aging can only exist in special circumstances (e.g. salmon) specifically excluding mammal aging.

Modern programmed aging theories are based on more recent mechanics theories to the effect that non-individual benefit can offset individual disadvantage and lead to the conclusion that purposely limited lifespan has a general evolutionary benefit specifically including mammals.

Theory Medical Implications

Programmed and non-programmed theories have very different predictions regarding the nature of mechanisms responsible for aging and age-related diseases and conditions.

Programmed aging theories suggest that the mechanisms responsible for diverse age-related diseases and conditions such as cancer and heart disease have common elements (e.g. the program, the biological clock, etc.). Pharmaceutical agents can therefore be sought that alter these common elements and result in simultaneously delaying or ameliorating multiple age-related diseases and conditions. *Anti-aging medicine* is therefore a realistic short-term possibility. *Regulated* programmed aging theories even more strongly suggest commonality and suggest existence of sensing and signaling mechanisms that offer additional targets for intervention.

Non-programmed theories suggest that aging is the result of deficiencies in a potentially large number of independent maintenance and repair mechanisms and that therefore there is little or no commonality between mechanisms responsible for the different age-related diseases and conditions. Medical research efforts consequently should be directed toward each disease or condition separately. Some non-programmed aging proponents (e.g. G. Williams) consider anti-aging medicine to be theoretically impossible based on their non-programmed theories.

Special Journal Issue – Programmed vs Non-programmed Aging 2009

In December 2008, eighteen scientists active in the field of biological aging theory received an interesting invitation. The invitation was from the Russian Journal of General Chemistry (RCJ) and requested articles for a special issue to be devoted to a 150-year-old but still unresolved question: Is aging in humans and other mammals "programmed" or "non-programmed?" That is, are the deteriorative and ultimately fatal effects of aging a purposeful result of the organism's genetically programmed design? Other commonly used ways of stating this same question: Is aging an *adaptation* in that the deteriorative effects serve some purpose and therefore are the result of evolved organism design features? Is aging the result of an *active*, pro-aging mechanism in the organism's design or is aging entirely the *passive* result of forces acting *upon* the organism? Does an organism's design fight aging, *cause* (or purposely allow) aging, or do both at different times in the organism's life (as proposed here)? Scientists representing both sides of the "programmed" argument were invited to participate in the project.

I submitted a pro-programmed article titled *The Case for Programmed Mammal Aging*[48] in response to the invitation. Because the special issue is read by a diverse audience the article avoided arcane terminology and included substantial background material. The request specified rather long articles and the journal is published in both Russian and English.

The RCJ special issue on aging is a major step toward increasing general scientific awareness of the programmed vs. non-programmed problem and also allows a wide audience to examine a side-by-side comparison of logical arguments by multiple representatives of both factions. This could provide major impetus toward finally resolving an issue with potentially major medical and public health implications.

Homo Sapiens Liberatus Workshop, Moscow 2010

In May 2010, a workshop titled *Homo Sapiens Liberatus* was conducted at Moscow State University. One of the assigned tasks of this workshop was to develop arguments regarding the evolutionary mechanics whereby an individually-adverse organism design characteristic having a non-individual benefit (e.g. evolvability or group-benefitting trait) could nevertheless be evolved and retained in an organism design. I presented two arguments, either of which, if correct, would support various programmed aging theories[49]:

The ***linkage argument*** suggests that linkage between an individually beneficial trait and an individually adverse trait could protect the individually adverse trait from being selected out. "Linkage" means the two traits are linked in such a way as to make it difficult for the evolution process to separate them and result in retention of the individually beneficial effect without the adverse side effect. This idea was originally proposed by G. Williams in support of his 1957 antagonistic pleiotropy theory of (non-programmed) aging. The argument was now made that linkage works better for programmed aging because it only has to withstand the evolutionary process (acting to separate benefit from adverse side effect) for a relatively short time as opposed to indefinitely in the case of Williams' proposal. This linkage allows a long-term group benefit to trade off against a short-term individual cost and supports group selection and evolvability considered as a group benefit.

According to a sort of beneficiary or *cui bono* analysis, evolvability traits benefit the group, species, or even future species and therefore opponents conflate an evolvability benefit with long-term group benefit and some use the same arguments they historically used against group selection, e.g. long-term benefit cannot offset short-term cost, long-term benefit is weaker and slower, etc. The ***evolvability argument*** suggests that the propagation mechanics of evolvability are completely different from group selection. Where group selection is trading a future group benefit against a current individual disadvantage, an evolvability trait contributes to a *precondition* that is necessary for the evolution process to operate and consequently is effective on the same time-scale as natural selection. Therefore an evolvability benefit can trade off against an individual disadvantage. This allows evolution of programmed aging based on evolvability benefit. See more extensive discussion in the evolvability section.

I have not seen specific counter-arguments against either of these arguments, either of which support generally applicable programmed aging.

Tom Kirkwood vs. Programmed Aging in 2011

Tom Kirkwood, author of the 1975 non-programmed disposable soma theory, is predictably a vocal opponent of programmed aging theories. By 2011, programmed theories of aging had acquired a significant following in the scientific community and Kirkwood felt a need to make formal arguments against programmed aging as opposed to merely denouncing such theories without any scientific rationale for doing so (e.g. Hayflick in 2004). In 2011, Kirkwood published a paper *On the programmed/ non-programmed nature of aging within the life history*[50] (with Simon Melov) attacking programmed aging theories and their authors and citing papers written by me and V. Skulachev among others.

We in the programmed aging "school" were rather pleasantly surprised. Their paper makes the case that mammal programmed aging is not supported by traditional individual-benefit-only evolutionary mechanics theory, a position that most adaptive programmed aging proponents already concede. Although it notes the existence of group and evolvability (non-individual benefit) theories it makes no specific arguments against them and even concedes they could be applicable in certain circumstances. Nor does it make arguments against any of the specific mammal programmed aging theories based on group selection or evolvability.

Even more favorably, their paper concludes that, considering only individual benefit, the net evolutionary disadvantage of aging must be "effectively zero" beyond some species-specific age. They have therefore set up a situation in which they have to argue that the evolutionary advantage of *limiting* life beyond that critical age cannot even minutely exceed the zero disadvantage of death and result in the evolution of a lifespan limiting mechanism, a very difficult case that they do not even attempt to make.

Finally, they tacitly concede that empirical evidence favors programmed aging. Instead, they contend that *interpretation* of empirical evidence should be biased in favor of non-programmed aging because of *their* evolutionary mechanics concepts.

I wrote a paper *On the programmed/ non-programmed aging controversy*[51] responding to the Kirkwood-Melov paper. Vladimir Skulachev also published a responding paper *Aging as a particular case of phenoptosis, the programmed death of an organism (A response to Kirkwood and Melov "On the programmed/non-programmed nature of ageing within the life history")*[52].

Proponents of programmed aging considered this exchange a definitive "win" for their side. Serious students of aging theory should read these three papers and also see *Aging by Design*[53] for a longer discussion of this exchange. See the following section for links to the online articles.

Biochemistry (Moscow) Phenoptosis

Vladimir Skulachev has initiated publication of a journal dedicated to aging and especially programmed aging called *Phenoptosis* as a periodic special issue of the existing journal *Biochemistry (Moscow)* published in English and Russian. Skulachev is chief editor of *Biochemistry (Moscow)*, which is widely indexed. *Phenoptosis* means programmed organism death as an extension of the term *apoptosis* or programmed cell death. The premier issue of *Phenoptosis* was published in 2012. See next section for free full-text access to this issue.

Recent Scientific Opposition to Programmed Aging Theories

> *"The way evolution works makes it impossible for us to possess genes that are specifically designed to cause physiological decline with age or to control how long we live." (Olshansky, Hayflick, and Carnes, Scientific American, 2004)*

At this point a reader might have an obvious question: What do proponents of passive (non-adaptive, non-programmed) aging theories currently have to say when presented with active (adaptive, programmed) theories? In recent years, I have had extensive interactions with many, sometimes very senior, proponents of passive, non-adaptive, non-programmed aging theories. Here are the four main lines of reasoning they presented as supporting their view that aging in humans is the result of passive non-programmed aging. I have not used their names because the following position statements are composites of several scientists with similar positions and/or contain material from private communications.

Evolution makes it impossible: Scientist "A" accepts and promotes the view quoted above to the effect that orthodox evolutionary mechanics theory makes active programmed aging "impossible" regardless of any amount of purported supporting observational evidence. He suggests that as a very outspoken and longtime proponent of passive aging, he would have to reconsider his position only if and when a single new programmed-aging-compatible evolutionary mechanics theory was to become "generally accepted." He is secure in the knowledge that this will almost certainly not occur, to his satisfaction, in his lifetime. In the meantime he suggests that I am trying to single-handedly overthrow 150 years of scientific tradition. He ostentatiously ignores the existence of the other programmed aging proponents as well as all those having issues with orthodox mechanics and multiple published non-individual mechanics theories. He further suggests that I might be some sort of closet Intelligent Design proponent merely for voicing any uncertainty regarding the absolute truth of orthodox evolutionary mechanics!

"A" is representative of people who developed or acquired their aging theory concepts during a time when it was indeed essentially universally considered that "evolution makes it impossible" for active aging to exist – i.e. theories and theorists descended from 1950s. "A" bases his position entirely on evolution theory considerations and expresses the most common position taken by non-programmed aging adherents. They ignore both observational evidence and developments that affect evolutionary mechanics theory. Some merely ignore proponents of programmed theories in the hope (so far futile) that they will eventually go away. Others write articles criticizing programmed aging theorists based on the given that evolution makes it impossible. "A" has a large investment in non-programmed theory. For "A" to change his mind on aging would be like an Episcopal bishop deciding to start over as an entry-level Methodist seminarian.

People who believe that it is *impossible* that orthodox mechanics theory could be less than perfectly comprehensive will also understandably believe *extremely improbable* orthodox explanations for apparently conflicting observations.

Evolution makes it functionally impossible: Scientist "B" accepts that there are issues with evolvability, group selection, and so forth that affect evolution theory on some intellectual level but contends that it is functionally impossible for these factors to affect evolved mammal aging mechanisms. He cites analyses (such as that performed by George Williams in 1966) that in his opinion "prove" that group benefits are too "weak" and "late" to override any individual disadvantage. He contends that evolvability benefits are "group" benefits (and therefore covered by the earlier analyses) despite logical arguments to the contrary. Because Medawar's and Williams' hypothesis (accepted by "B" despite its population-oriented nature) says that the net individual disadvantage of aging is negligible, he is essentially taking the position that there is no evolvability or group benefit that is so large that it could override the smallest possible individual disadvantage. Keep in mind that Williams was the author of a non-programmed aging theory and therefore not exactly an impartial analyst. "B" does not offer any thoughts as to how all those apparently individually adverse characteristics came to be evolved and retained. "B" also avoids discussing either the details of current evolutionary mechanics thinking or the details of current observational evidence supporting active aging.

"B" acknowledges the existence of alternative mechanics theories while denying that they have any operational validity whatsoever. "B" likely believes that this path is superior to "A's" "head-in-the-sand", "pretend-it-doesn't-exist" approach to the problem of a changing evolutionary mechanics landscape.

Non-programmed aging is functionally identical to programmed aging: Scientist "C" (amazingly similar to Aubrey de Grey) accepts that there are issues with evolutionary mechanics theory but contends that a passive mechanism based on Medawar's hypothesis would be functionally identical to an active lifespan limiting mechanism. They both result in mammal lifespans that vary with puberty age and other species-specific factors. If this was indeed true, there would be no selectable property differentiating a passive mechanism from an active mechanism regardless of the evolutionary mechanics theory chosen.

"C" says aging results from an organism's inability to better resist universal deteriorative processes. He contends that this passive mechanism is essentially the "default" given the existence of universal deteriorative factors such as oxidation, wear, and molecular damage and suggests that there would therefore be no evolutionary force leading to the development, selection, and retention of a more complex active aging mechanism.

He accepts that some species (semelparous species including semelparous mammals) possess active suicide mechanisms but says they are "irrelevant" to gradual mammal aging. He suggests (in my opinion) wildly implausible passive-compatible (and orthodox-compatible) explanations for various other observations described in this book that support active programmed aging over passive aging in mammals. For some of the observations he presents no passive-compatible explanation.

"C" disdains the idea that an evolved active aging mechanism has many functional advantages in flexibility and ability to adapt to temporary or local conditions (regulation) that would be useful to mammals and therefore would be selected. An active mechanism could be very complex and could be functionally very different from his suggested passive mechanism. Observational evidence supports the functionally more complex case.

Further, if one accepts that deterioration at a certain point in an animal's life provides benefits then would not the obvious "default" be that the deteriorative processes are "programmed" by an extension of the same program that programmed growth, development and other major life events such as puberty?

Although "C" claims neutrality regarding the evolution controversies he uses many orthodox arguments and accepts as facts many orthodox explanations for apparently conflicting observations.

"C's" position was designed for and works well with a single observation: Mammal lifespans vary dramatically between species. "C's" main problem is that programmed theories provide a much better match than passive theories to a long list of *other* observations while also matching this observation.

Tradition, Seniority, and Popularity: Scientist "D" like most people in life sciences has been trained from an early age to believe in passive aging but is not the author of or otherwise heavily invested in passive theories and does not consider himself an expert regarding aging theory. He is aware of the turmoil that currently exists regarding evolutionary mechanics theory and does not believe evolution theory should be the overriding consideration in aging theory. He also does not have logical counter-arguments regarding the observational evidence supporting active aging. Nevertheless he still believes in non-programmed aging. "D's" views are common in the general bioscience community. We may hope that he is now having some doubts.

"D's" is not an unreasonable position. We are all familiar with the mathematical parlor tricks that purport to prove that 22 is equal to 23 or other equally unlikely proposition. Just because one cannot personally figure out the counter-argument (within the time and effort one is willing to invest) does not mean that 22 is equal to 23. Somewhere there is a person or group that does have the competence, time, and energy to produce logical arguments as to how the trick was performed and why the conclusion is erroneous. If I am very sure of my position, (and I am indeed extremely sure that 22 does not equal 23) I will not seek such a confirmation. It is not unreasonable for a non-expert to adopt the scientifically popular view or to follow the thinking of experts in the community.

Resolving the Programmed vs. Non-Programmed Aging Controversy

Notice that the positions described above express progressively less certainty regarding evolutionary mechanics. Developers and others heavily invested in passive theories such as "A" tend to also be very sure that orthodox evolutionary mechanics theory is correct, complete, and totally comprehensive and reject even the slightest possibility that this might not be the case. Those like "D" who are less centrally involved in aging theory tend to also be much less committed to orthodox mechanics theory.

My analysis of the aging theory situation is as follows:

Orthodox evolutionary mechanics theory mandates passive non-programmed aging. Proponents of passive aging emphasize evolution considerations and deemphasize observations, often to an extreme extent.

Current observational evidence strongly favors active programmed aging.

Alternatives to orthodox mechanics theory, primarily developed *after* origination of the most popular passive aging theories, appeared in efforts to explain observational discrepancies in areas *other* than aging. Several of these alternatives (group theories, evolvability theory) allow programmed aging, support the need for programmed aging, and provide predictions regarding the nature of programmed aging.

Proponents of non-programmed aging currently are more numerous and also tend to be more senior and have more influence in terms of academic pecking order than proponents of programmed aging.

Most bioscientists accept non-programmed aging because of tradition, popularity, and seniority but also tend to accept that orthodox evolutionary mechanics has serious issues based on current science. This represents a logical "disconnect." Many of these people would have second thoughts about aging if they understood the degree to which passive theories depended on orthodox evolutionary mechanics.

Non-programmed proponents "B" and "C" above have made efforts to accommodate the changing evolutionary mechanics situation and counter programmed proponents without violating their core beliefs. In my opinion the arguments made by "B" and "C" would not withstand any serious unbiased scientific analysis for the reasons given.

Medical people (not to mention medical patients) tend to be extremely pragmatic. If it works, they do not care about the theory. It now appears possible that agents with general anti-aging properties will be accidentally discovered and exploited. This would lead to searches for additional agents and tests of agents with similar chemical structures. Subsequent analysis will eventually determine mechanisms and improved agents based on that knowledge will be developed. If this empirical scenario develops rapidly, academic arguments about aging theory may eventually be moot.

However, this approach might seriously delay development and availability of anti-aging agents. If experimentalists follow only orthodox theories, they might be looking in the wrong places. Some suggest that medical science regarding age-related conditions has already been delayed for decades by orthodox aging theories. Therefore resolving these issues is not actually an "academic" problem but rather a problem that could dramatically affect medical science and therefore general health and welfare.

In the theory world we have a situation in which some theorists are pursuing aging theories based on the absolute truth and comprehensiveness of orthodox evolutionary mechanics theory while simultaneously others are trying to modify orthodox theory in order to handle the issues described earlier. It is obvious that both approaches cannot be simultaneously correct. In an ideal world, we would first definitively solve the evolution theory problems and then solve the dependent aging theory problem. As a complicating factor, the evolution issue is highly related to the aging issue in that observations about aging impact evolution theory in addition to the converse.

The aging theory situation could be resolved by a scientific task force consisting of unbiased but senior and highly respected scientists who would review current observational evidence and current evolutionary mechanics thinking, listen to proponents from both sides, and produce a conclusion regarding aging theory. The conclusion could find for passive or active aging and provide a logical rationale or could also determine that additional investigation was needed and suggest the nature of that investigation. Needless to say, I am confident that such a task force would find in favor of programmed aging. The difficulty would be in finding unbiased senior scientists to staff the review panel.

The task currently confronting programmed aging proponents is therefore to create enough doubt in enough people that such a "zero-base" analysis process will indeed be sought and conducted regarding the aging theory issue. Progress is being made in this direction. Gerontology journals are (2014) increasingly willing to accept pro-programmed aging articles.

Summary Situation

There are a number of viewpoints or perspectives that bear on the question: "Why do we age?" and suggest radically different conclusions.

The ***human perspective*** on aging, including our accumulated medical knowledge of the effects of aging, strongly suggests that aging is a sort of generic, unavoidable, and accumulative degradation similar to that which occurs in non-biological systems. We use the same word, "aging" to describe such deterioration in humans, machinery, buildings, or exterior paint. Because the human experience is so ubiquitous and has such a major impact on people's lives, it has a profound effect on popular and medical thinking about aging and doubtless influences other scientific thought.

The ***biology perspective***, resulting from a comparative analysis of lifespan characteristics of a wide variety of different organisms including mammals, reptiles, fish, birds, insects, and even plants, presents an entirely different picture. This viewpoint leads to the essentially unavoidable conclusion that lifespan control is a part of an organism's design. Some species have extremely explicit mechanisms for lifespan regulation.

Darwin's theory of evolutionary mechanics, itself based on observations, holds that *individual* short-term survival and reproduction is the main driver of evolution. According to orthodox theory, it is impossible for an organism to evolve a characteristic that limits its lifespan unless that characteristic simultaneously promotes reproduction.

The ***traditional theories of aging***, developed between 1950 and 1980, provided a semi-plausible reconciliation of the dilemma formed by Darwin's theory and the comparative lifespan analysis while neatly confirming the human experience. Although they conflicted with each other, ignored the lower species, and had many logical flaws it was reasonable to believe that eventually a single unified theory would be developed that successfully reconciled the observational evidence without violating "orthodox Darwinism."

However, such a unified theory has not appeared. Instead, more ***recent discoveries*** such as aging genes and details of genetics, shed doubt on not only the traditional theories but also some of the details of orthodox Darwinism. Digital genetics anomalies and behavioral discrepancies also suggest that adjustment of Darwin's theory is necessary. At least two such proposed adjustments allow lifespan regulation to be an adaptive evolved characteristic, a purposeful part of an organism's design.

Research Inhibiting Factors

Anti-aging research has been and continues to be inhibited relative to research regarding other public health issues, conditions, and diseases, by a number of factors in addition to the fact that very little clinically demonstrated progress has been made:

Perception that aging is unalterable. The great majority of the public considers aging to be an essentially unalterable fact of life and that therefore significant medical intervention in aging is impossible or very improbable. Anti-aging research is therefore seen as foolish and wasteful.

This opinion is not based on any particular scientific theory and is probably mainly a result of the perception that aging is universal and that it more or less uniformly affects all living things. People who are aware of evidence that this is not the case such as progeria, Werner's syndrome, "non-aging" species, and details of inter-species variations, or who are aware of evidence that aging can be contravened such as experiments with caloric restriction and aging genes, tend to be much more favorably inclined regarding anti-aging research.

Perception that aging is "normal." A minority of the public and probably a significant number of physicians think of aging as "normal" and therefore not necessarily a proper subject for medical intervention. Heart disease, cancer, and other "age-related" diseases, while collectively a "normal" consequence of aging, are individually not "normal" and therefore legitimate targets of medicine. This logic becomes less supportable if people come to believe that aging predisposes and therefore is a substantial "cause" of age-related disease, and that therefore "anti-aging research" is attacking the age-related diseases. Many pharmaceuticals such as vaccines and painkillers are intended to alter the "normal" design of the human organism.

Outmoded scientific theories. Prevailing, widely taught, and highly publicized traditional scientific theories of aging dating from the 1950s are very pessimistic regarding the potential for major medical intervention in aging. These theories are unproved, have major flaws, and are specifically incompatible with some more recent discoveries. Newer, at least equally plausible adaptive theories are much more optimistic.

Because of an almost religious belief in un-testable and un-provable aspects of evolution theory, many scientists have been led to ignore much more direct and experimentally demonstrable evidence.

Experimental difficulty. Because of the long-term nature of aging, anti-aging research is experimentally difficult compared to most disease research. Experiments, critical to any medical advance, are difficult to design and perform, time consuming, and expensive. It is obvious that additional funding would greatly accelerate the pace of anti-aging research and would allow more parallel as opposed to serial experiments to be performed.

Scams and Quacks. Legions of scammers and quacks selling "anti-aging" goods and services lead to an atmosphere in which any anti-aging effort is seen as suspect.

Non-aging species, caloric restriction experiments, research on mimetics, discovery of aging genes, Werner's syndrome, and discoveries relating aging to hormones all strongly suggest that aging is a potentially highly treatable condition. A major treatment for aging might not be much more difficult than a major treatment for AIDS.

One reason for being optimistic about anti-aging prospects is that aging, considered as a disease, progresses very slowly and therefore even a relatively minor interference in the aging process would have a major beneficial effect on public health.

A treatment resulting in a 50 percent improvement in post-symptom-onset lifespan for rabies patients would not be considered very significant since it would only extend life by a short time and besides, only a very few people contract rabies. To be effective, a treatment would have to essentially produce a "cure", a complete elimination of the disease's effects. Similarly, a 50 percent improvement in post-onset survival of lung cancer patients or even AIDS patients may not be considered a major breakthrough.

However, a 50 percent increase in post-onset survival of "aging disease" which eventually affects most of the population and could result in an average life extension of 20 years or more would have massive positive public health impact. Once an aging mechanism has been completely identified, producing such a relatively minor interference with the mechanism should be within reach. In the case of aging, a cure is not necessary to produce a major improvement in public health. Early research on aging genes and caloric restriction has already produced interference at or more than the 50 percent level in various organisms.

Furthermore, insight into the aging process would almost certainly lead to better understanding of age-related conditions such as heart disease, cancer, stroke, and arthritis that will eventually affect almost every person. To illustrate, extensive and well-funded research on heart disease has been underway for at least 75 years. To at least some extent a point of diminishing return has been reached. Anti-aging research might well provide a new approach or a "new angle" to the treatment of heart disease as well as other age-related diseases and conditions.

Observational evidence and theory suggesting that aging is biologically controlled in a manner similar to the control of other features such as puberty age is especially interesting. Most scientists would agree that it would not be medically difficult to find a way to advance or retard age of puberty. It is possible that a similar approach could eventually be applied to aging.

Prevailing scientific attitudes are based largely on traditional theories of aging dating from the 1950s and *not* on the discoveries listed above. Present and future medical researchers and physicians have been and are still being taught theories leading to the conclusion that significant medical intervention in aging is unlikely or impossible.

Historical Summary of Evolutionary Aging Theories

Here is a brief summary of aging theory development.

It was apparent that the observed extremely species-unique nature of lifespan required a theory that allowed for lifespan to be either an organism design characteristic or, at a minimum, very dependent on other species-unique design characteristics. This invoked evolutionary mechanics theory in that such a theory explains how organisms acquire their species-unique designs.

In 1952 there was essentially no serious scientific opposition to orthodox evolutionary mechanics theory. Given this constraint and within a multi-species context, Medawar's 1952 hypothesis appeared very attractive as an evolutionary explanation for gradual aging in mammals. However, since then, there have been many developments and non-developments that dramatically change that picture regarding both aging theory and evolution mechanics theory.

Despite passage of a half-century, there has not been any convergence regarding an aging theory based on Medawar's hypothesis or otherwise based on orthodox mechanics. Instead multiple competing theories still exist. Each succeeding theorist criticized logical flaws in the proceeding theory and others have subsequently found major flaws in all of them. Further, experimental confirmation of related predictions (e.g. demonstrated rigid link between aging and some individually beneficial function) has not occurred.

Contemporaneously, additional experimental evidence supporting existence of lifespan management systems (e.g. aging genes, hormone mediated aging mechanisms, etc.) has been found.

Although scientific certainty regarding the validity of Darwin's species-origin ideas continues today, there has been substantial divergence concerning orthodox evolutionary mechanics theory. Multiple proposals for adjustments or corrections to orthodox mechanics now exist including group selection (1962), kin selection (1963), selfish gene theory (1975) and evolvability theory (~1996).

Major issues now exist regarding orthodox mechanics in at least three different areas:

Observational discrepancies: A list of different categories of observed organism design characteristics that appear to be incompatible with orthodox theory (Chapter 2) remains 150 years after Darwin. There are thousands of observations of organism design characteristics that appear to be individually adverse or neutral. The observations are not generally in dispute but rather the interpretation or explanation of those observations varies. Orthodox explanations generally claim that the apparently adverse design characteristic has some compensating individual benefit. Some orthodox explanations are entirely circular and claim the existence of an unspecified (in addition to undemonstrated) individual benefit.

Evolvability: Orthodox theory assumes that all organisms have the same capacity for evolution, a position that is now clearly false. Any valid evolution mechanics theory must therefore deal with the evolvability issue.

Evolutionary Effect of Inheritance Mechanisms: Orthodox theory assumes that natural selection is the *only* factor that *differentially* affects propagation of mutational changes. However, discoveries in genetics science (many since 1952) have disclosed design features in sexually reproducing species that clearly do differentially affect propagation of a mutational change based on many complex and interacting factors. A valid evolutionary mechanics theory must deal with this issue. There are similarly many journal articles that discuss aspects of this problem. The additional complexity introduced by this issue increases the plausibility of group selection and kin selection and therefore the plausibility of aging theories based on these alternatives.

Goldsmith

The following figure provides a chronological sequence showing dates at which major evolutionary mechanics concepts were introduced and consequent modern non-programmed and programmed aging theories.

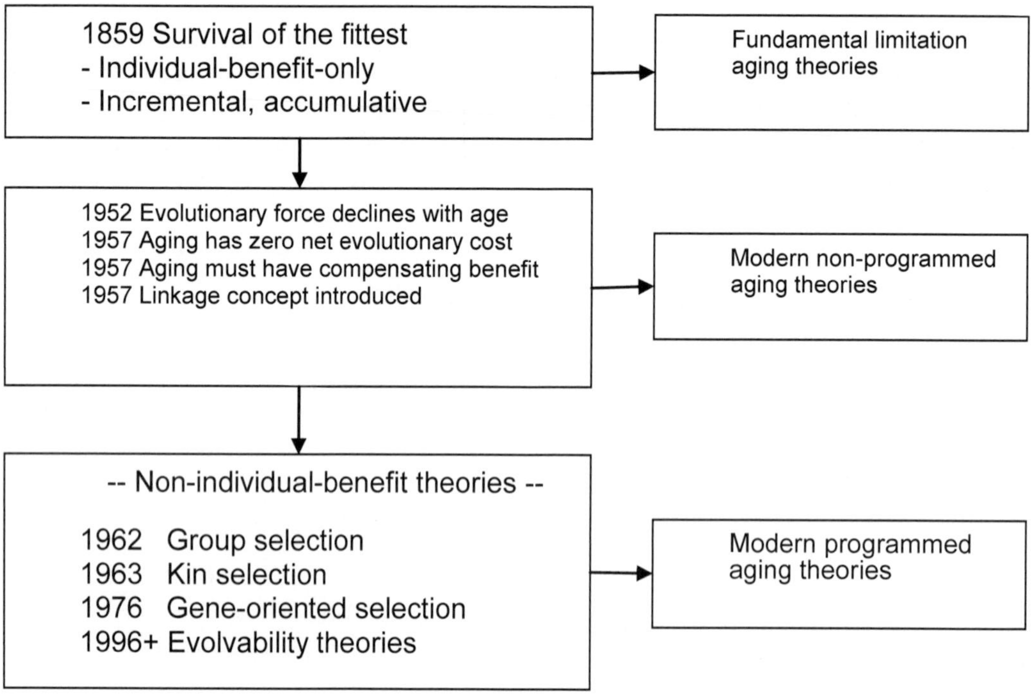

Figure 29 Timeline of Major Evolutionary Mechanics Concepts

12. Online Resources

The author maintains an Internet web site on aging at: http://www.azinet.com/aging/ including links to many on-line resources. Many resources regarding evolutionary theory and other aspects of aging are available on the Internet. When an Internet resource has been identified it is given in the references below. A clickable version of this list is available on the web site if you would like to access the on-line material without having to type the long URLs.

The author may be contacted at tgoldsmith@azinet.com.

Programmed-aging.org

The author and several other theorists who are proponents of programmed aging maintain a web site at: http://www.programmed-aging.org/ that provides information on programmed (adaptive) aging and the programmed vs. non-programmed aging controversy. This site provides access to information suitable for a wide audience.

Programmed/ Non-Programmed Aging Controversy 2011-2014

Against Programmed Aging:

T. Kirkwood and S. Melov *On the programmed. Non-programmed nature of ageing within the life history*. Current Biology 21, R701–R707, September 27, 2011 DOI 10.1016/j.cub.2011.07.020.
Full text PDF access:
http://ac.els-cdn.com/S0960982211007895/1-s2.0-S0960982211007895-main.pdf?_tid=c058c87c-4938-11e2-ba83-00000aacb360&acdnat=1355852066_f36650bbda025137c72a1e318ec96f4e

For Programmed aging:

T Goldsmith. *On the programmed/ non-programmed aging controversy*. Biochemistry (Moscow) Phenoptosis, Vol 77 No 7, pp. 729_732, 2012 DOI: 10.1134/S000629791207005X
Full text PDF access:
http://www.azinet.com/aging/programmed_vs_nonprogrammed.pdf

V. Skulachev. *Aging as a particular case of phenoptosis, the programmed death of an organism (A response to Kirkwood and Melov "On the programmed/non-programmed nature of ageing within the life history")* Aging V3(11) 2011
Full text PDF access:
http://www.impactaging.com/papers/v3/n11/pdf/100403.pdf

T Goldsmith. *Arguments against non-programmed aging theories*. Biochemistry (Moscow) Phenoptosis, 78:9 971-978 2013

Goldsmith

http://www.azinet.com/aging/non-programmed_aging.pdf

T Goldsmith. *Aging Theories and the Zero-Sum Game*. Rejuvenation Research. February 2014, 17(1): 1-2. doi:10.1089/rej.2014.1548 (This guest editorial in Rejuvenation Research suggests approaches to resolving the programmed/ non-programmed aging controversy.)

http://www.azinet.com/aging/Zero_sum_game.pdf

http://online.liebertpub.com/doi/abs/10.1089/rej.2014.1548?journalCode=rej

Biochemistry (Moscow) Phenoptosis

Premier issue 2012 Vol 77(7) Full-text access at:
http://protein.bio.msu.ru/biokhimiya/contents/v77/ToC7707.html

2013 issue Vol 78(9)
http://www.protein.bio.msu.ru/biokhimiya/contents/v78/ToC7809.html

Darwin

The Origin of Species - Charles Darwin, 1859
http://www.literature.org/authors/darwin-charles/the-origin-of-species/

The Descent of Man – Charles Darwin, 1871
http://www.literature.org/authors/darwin-charles/the-descent-of-man/

Non-programmed Theories of Aging

An Unsolved Problem of Biology Medawar, P.B., 1952. H.K. Lewis & Co., London. http://www.telomere.org/Downloads/Medawar-UPB.pdf Medawar's paper provided a proposed model for the characteristics of a non-aging species and introduced the mutation accumulation theory of aging.

Pleiotropy, natural selection and the evolution of senescence, Williams, G. 1957. *Evolution* 11, 398-411 http://www.telomere.org/Downloads/Williams_searchable.pdf Williams' antagonistic pleiotropy theory is one of the most respected of the traditional theories.

Senescence.info - A collection of information and links on aging, and gerontology by Joao Magalhaes at Harvard: http://www.senescence.info/ Magalhaes is a proponent of non-programmed aging.

No Truth to the Fountain of Youth - Olshansky, Hayflick, and Carnes, *Scientific American* June 2002 (reprinted July 2004 Vol 14. No. 3) – This article provides warnings against common ineffective anti-aging remedies. Aging is an "inescapable biological reality" caused by the accumulation of random damage to the building blocks of life. Fifty-one traditional scientists endorsed this relatively recent position paper to the effect that aging is not and cannot be an evolved adaptation.

http://www.sciam.com/explorations/2002/051302aging/

Evolutionary Theories of Aging and Longevity - L. A. Gavrilov et al; University of Chicago Center on Aging -- Links to many articles by this team - Excellent overview of "evolutionary" theories of aging including Weismann (adaptive) and mutation accumulation / antagonistic pleiotropy (non-adaptive). Describes negative impact of some aging theories on research - Cautions that all aging theories are just theories and should not unduly influence research. http://www.src.uchicago.edu/~gavr1/

New Adaptive Theories of Aging

Aging is a Specific Biological Function Rather than the Result of a Disorder in Complex Living Systems: Biochemical Evidence in Support of Weismann's Hypothesis, V. P. Skulachev Moscow State University -- http://www.thehormoneshop.net/telomeraserussian.htm

Aging selected for its own sake Joshua Mitteldorf *Evolutionary Ecology Research, 2004*, **6**: 1 – 17 Temple University Provides an extensive compendium of experimental evidence against traditional theories especially antagonistic pleiotropy and disposable soma theories. http://www.mathforum.org/~josh/4OwnSake.pdf

Whence Cometh Death J. Mitteldorf; University of Pennsylvania – This site contains a good discussion of group selection and the evolved vs. non-evolved controversy as well as links to most of Mitteldorf's articles. http://mathforum.org/~josh/

Aging as an Evolved Characteristic – Weismann's Theory Reconsidered Theodore. C. Goldsmith *Medical Hypotheses* 2004 62-2 304:308 DOI: 10.1016 S0306-9877(03) 00337-2. Article discusses evolutionary disadvantages of immortality and evolvability theory of aging. http://www.azinet.com/aging/aging-theory3.pdf

Regulation of Life-Span by Germ-Line Stem Cells in Caenorhabditis elegans, Cynthia Kenyon, *Science* (Vol. 295, 18 January 2002). Free registration required for full-length article access at: http://www.sciencemag.org.

An Introduction to Biological Aging Theory (PDF) T. Goldsmith Rev 2012. ISBN: 0-9788709-1-3. Apple ibook Kindle ebook Nook ebook Sony ebook Kobo ebook

Introductory text describes the main theories of aging.

Aging by Design (PDF) T. Goldsmith Rev 2012. ISBN: 0-9788709-3-X. Kindle ebook Apple ibook Nook ebook Sony ebook Kobo ebook

Short book written for a wide audience describing the history of aging theories and theorists. Makes the case for programmed aging.

New Truth to the Fountain of Youth (PDF) T. Goldsmith 2012 ISBN: 0-9788709-4-8. Nook ebook Kindle ebook Apple ibook Sony ebook Kobo ebook

Goldsmith

Short book describes medical implications of various aging theories and suggests ways of using the Internet and other modern technology to find anti-aging agents.

Anti-Aging Research

Daf-2, an insulin receptor-like gene that regulates longevity and diapause in Caenorhabditis elegans. Kimura KK, Tissenbaum HA, Liu Y, Ruvkun G. *Science* 1997; 277: 942. A report of the discovery of a gene that controls aging in the roundworm. http://www.sciencemag.org.

An Engineer's Approach to the Development of Real Anti-Aging Medicine Aubrey D.N.J. de Grey *Sci Aging Knowledge Environ.* 2003 Jan 8;2003(1):VP1. http://www.gen.cam.ac.uk/sens/manu16.pdf

Other Resources

Progeria Research Foundation http://www.progeriaresearch.org/

Werner's Syndrome Overview http://depts.washington.edu/statgen/Computing/wsbackgrnd.html

AgelessAnimals Information on Long-lived Animals with Negligible Senescence: http://www.agelessanimals.org

Longevity Records: Lifespans of Mammals, Birds, Amphibians, Reptiles, and Fish, Max Plank Institute, ISBN 87-7838-539-3 -- The oldest lake sturgeon caught so far was 152 years old. *http://www.demogr.mpg.de/longevityrecords/*

DesertUSA Information on Bighorn Sheep: http://www.desertusa.com/big.html DesertUSA.com

Human Genome Project. The HGP is an approximately $3 billion government effort to fully sequence the human genetic code. The effort, which began in 1990 was substantially completed in 2003. The second link is for online copies of the actual preliminary project reports dated 2001.

http://www.ornl.gov/TechResources/Human_Genome/
http://www.ornl.gov/TechResources/Human_Genome/project/journals/journals.html

Age-Specific Survival In Five Populations Of Ungulates: Evidence Of Senescence Anne Loison, et al, *Ecology,* 80(8), 1999, pp. 2539–2554. This study followed large wild animals including Bighorn Sheep to determine if aging in wild animals, in fact, significantly altered death rates. Results show death rates indeed significantly increased with age contrary to traditional aging theory.

http://www.callisto.si.usherb.ca:8080/caprinae/pdffiles/Loison_et_al.pdf

NCBI Bookshelf National Center for Biotechnology Information collection of free on-line searchable books. NCBI is part of the NIH National Library of Medicine. Major subjects include: Biochemistry, Endocrinology, Genomes, Genes and Disease, Immunobiology, Molecular Biology, and Medical Microbiology http://www.ncbi.nlm.nih.gov/entrez/query.fcgi?db=Books

From the Publisher

Please take a moment to rate this book. If you liked or disliked something about this book, the publisher, author, and future readers would very much appreciate it if you would also write a review.

Comments, suggestions, or inquiries for the publisher may be sent to: books@azinet.com.

--Azinet Press

Appendix – Aging Attitudes Survey

The survey of aging attitudes and knowledge was conducted using the search service "SeekOn" (http://www.seekon.com/) in early 2003. SeekOn provides local what-to-do and where-to-stay information for visitors and residents of about 15,000 towns in the U.S. and Canada and therefore attracts family-oriented or business-oriented users. Users are more highly educated, older, and more likely to be female than average web browsers.

The survey is not rigorously scientific because participants selected themselves. People who desired to take such a survey might be predisposed to answer certain questions in a particular way. However, the results of demographic questions (age, sex, etc.) closely follow demographics for the site measured independently. These results are therefore considered highly indicative of attitudes and knowledge in the general population.

The survey was presented as four sequential pages that allowed measuring responses to a question without contaminating the respondent by exposing the subsequent questions. Each question was presented as a multiple choice in which the respondent could decline to answer or could check a single answer but (mechanically) could not select multiple answers. Respondents could not change their answers on previous pages once they moved to a subsequent page. Not all respondents answered every question; the results listed below are percentages of the answers given for the question indicated.

The Survey questions and the answers recorded from the 142 respondents are listed below:

> SeekOn is supporting a research project to determine attitudes regarding and knowledge of the aging process. Any information you supply in this poll will only be used for statistics. No personally identifiable information will be acquired. At the end of the questionnaire you will be able to see the results (so far) of the poll. You should find it interesting.
>
> Thanks for your support of this project.
>
> **O** Do you live in: [82%] United States [15%] Canada [3%] Other
>
> **O** Your Age: [9%] Under 20 [23%] 21 - 30 [23%] 31 - 45 [21%] 46 - 55 [17%] 56 - 65 [7%] Over 65
>
> **O** Your Sex: [48%] Male [52%] Female
>
> **O** Your Education: [30%] High School [34%] Some College [25%] College Degree [11%] Graduate School
>
> **O** Have you ever studied biology? [19%] No [58%] Yes, High School Only [23%] Yes, College
>
> **O** What do you think is the most likely cause of aging?
>
> [28%] All living things eventually wear out.

[29%] Damage to cells, DNA, or other critical function gradually accumulates.

[36%] We are designed to age.

[7%] Nobody knows. We may never know.

O Which of the following most closely describes your views about anti-aging treatments?

[62%] Aging is an inescapable biological reality - There will never be meaningful treatment of the fundamental causes.

[18%] Some day in the very distant future they might find a treatment.

[6%] Treatment of the fundamental causes of aging is possible in the relatively near term.

[7%] A major treatment for aging might be as easy to do as a major treatment for AIDS.

[7%] Effective, significant, treatments are already available such as HGH.

The National Institute of Aging (NIA)(part of the U.S. National Institutes of Health(NIH)) provides funds to study fundamental causes of aging as well as study of some specific age related diseases such as Alzheimer's. In 2003 NIA's budget request was about $965 million. The study of AIDS was funded at $2.8 billion. Total NIH budget was about $27 billion. For comparison, expenditures for chewing gum in the U.S. are about $2 billion annually.

O Do you think taxpayer provided funding for fundamental research on aging should be: [32%] Increased [25%] Decreased [43%] Stay the Same

O Do you think anti-aging research has any moral issues? [43%] No [36%] I am somewhat concerned [20%] Yes, we should not try to extend natural lifespan.

O Did you know that there are species that apparently do not age such as yellow-eye rockfish and some turtles? [22%] Yes [78%] No

O Did you know that genes have been found in mice and other organisms that apparently cause aging? Inactivation of these genes through genetic engineering has extended average lifespans by as much as 50 percent. [30%] Yes [70%] No

O Did you know that restricting caloric intake of lab rats while maintaining a nutritious diet has extended average lifespans by as much as 50 percent? The rats are healthier in addition to living longer. Similar results have been observed in other animals. [52%] Yes [48%] No

O Did you know that researchers are searching for a medication that would mime the anti-aging effects of caloric restriction without having to actually restrict consumption? Preliminary results are encouraging. [23%] Yes [77%] No

O Did you know that the diseases causing the largest numbers of fatalities are all age related? Ninety percent of Americans who died in 1999 were over 57. [48%] Yes [52%] No

O If you want to leave a comment, please enter it in the box below. Be advised that your comment may be used (anonymously) in reports or articles.

INDEX

About the Author

During his more than 30 years at NASA's Goddard Space Flight Center, Theodore Goldsmith held many different positions mainly specializing in the design, development, and management

of digital data systems for NASA scientific spacecraft including the Space Shuttle, and the Hubble Space Telescope. He has been a computer programmer, digital systems engineer, microcircuit designer, and project manager and is a recipient of NASA's Exceptional Service Medal. Prior to joining NASA, Goldsmith worked for the National Institutes of Health.

In 1995 he became interested in the digital aspects of genetics and has written numerous articles and books about genetics, evolution theory, aging theory, and other technical subjects.

Goldsmith has a degree in electrical engineering from the Massachusetts Institute of Technology and is the owner of a small Internet company. He lives with his wife in Annapolis, Maryland.

Notes and References

[1] National Center for Health Statistics. *Vital Statistics of the United States*, Volume II: Mortality, Part A. Washington, D.C.: Government Printing Office, various years. (Data obtained through the Human Mortality Database, www.mortality.org)

[2] Charles Darwin, *On the Origin of Species*, 1859, ISBN 0-375-75146-7

[3] Weismann, August, *Uber die Dauer des Lebens*, Fischer, Jena

[4]Gavrilov, LA, The reliability-engineering approach to the problem of biological aging. Ann N Y Acad Sci. 2004 Jun;1019:509-12.

[5] Medawar, P.B, *An Unsolved Problem of Biology*., 1952. H.K. Lewis & Co., London.

[6] Williams, G *Pleiotropy, natural selection and the evolution of senescence,*. 1957. *Evolution* 11, 398-411

[7] Kirkwood T.B.L. & F.R.S. Holliday, *The evolution of ageing and longevity*, 1979. *Proceedings of the Royal Society of London B* 205: 531-546

[8] Lewin, Benjamin, Genes VIII, ISBN0-13-145140-5 Pearson Prentice Hall 2004

[9]Mandel, Gregor, Versuche über Pflanzen-Hybriden, 1865

[10] Watson, J, Crick, F. *A Structure for Deoxyribose Nucleic Acid, Nature*, April 1953

[11] Jasny, B.R.et al, *The Completion of the Human Genome Project, Science* April 11, 2003

[12] Cohn, E. *The Art of Genes*: *How Organisms Make Themselves*, 2000, Oxford Press ISBN: 0192862081

[13] Luckinbill LS, et al, Selection for lifespan in Drosophila melanogaster, Heredity. 1985 Aug;55 (Pt 1):9-18.

[14] Baret, P.V. et al, Selection for increased longevity in Drosophila melanogaster: reflections on new data.Gerontology. 1996;42(1):14-7

[15] Cockburn, A. *Lek promiscuity in a semelparous mammal,* Behavioral Biology and Sociobiology, 1988

[16] Bennett, J.T. et al. *Confirmation on longevity in Sebastes diploproa (Pisces: Scorpaenidae) from 210Pb/226Ra measurements in otoliths*. 1982. *Maritime Biology*. 71, 209-215.

[17] Buffenstein R. *The naked mole-rat: a new long-living model for human aging research*. J Gerontol 2005.

[18] Kenyon, C. *Regulation of Life-Span by Germ-Line Stem Cells in Caenorhabditis elegans,* , *Science* (Vol. 295, 18 January 2002)

[19] Apfeld J, Kenyon C. *Regulation of lifespan by sensory perception in Caenorhabditis elegans*. Nature 1999.

[20] Wolkow C et al. Regulation of C. elegans life-span by insulinlike signaling in the nervous system. Science 2000.

[21] Weindruch R, Walford RL, Fligiel S, Guthrie D. *The retardation of aging in mice by dietary restriction: longevity, cancer, immunity and lifetime energy intake*., *J Nutrition* 1986; 116: 641-54

[22] Lane, M. et al *The Serious Search for an Anti-Aging Pill Scientific American* Aug 2002 Describes experiments with 2DG to simulate the effect of Caloric Restriction.

[23] Wodinsky, J. 1977. *Hormonal inhibition of feeding and death in octopus: control by optic gland* secretion. Science, 198: 948–951.

[24] Smith, Frankel, and Yarnell, *Sex and death: are they related? Findings from the Caerphilly cohort study,* British Medical Journal, 1997

[25] Vito, P., E. Lacana, L. D'Aadmio Interfering with apoptosis: Ca2+ binding protein ALG-2 and Alzheimer's disease gene ALG-3, Science 1996:271:521-5

[26] Olshansky, Hayflick, and Carnes, *No Truth to the Fountain of Youth*, *Scientific American* June 2002 (reprinted July 2004 Vol 14. No. 3

[27] Loison, A. et al, Age-Specific Survival In Five Populations Of Ungulates: Evidence Of Senescence Ecology, 80(8), 1999, pp. 2539–2554

[28] Wayne-Edwards V. *Animal Dispersion in Relation to Social Behaviour*, Edinburgh: Oliver & Boyd, 1962

[29] Williams G. Adaptation and Natural Selection: A Critique of Some Current Evolutionary Thought, Princeton UP. ISBN 0-691-02357-3

[30] Wynne-Edwards, V. (1986) *Evolution Through Group Selection*, Blackwell. ISBN: 0-632-01541-1

[31] Hamilton, W, *The Evolution of Altruistic Behavior*, American Naturalist 97:354-356, 1963

[32] Mitteldorf, Joshua, *Aging selected for its own sake*, Evolutionary Ecology Research, 2004, **6**: 1 – 17

[33] Mittledorf, Joshua. *Chaotic Population Dynamics and the Evolution of Ageing*. Evolutionary Ecology Research 2006, 8: 561-574

[34] Travis J. The Evolution of Programmed Death in a Spatially Structured Population. Journal of Gerontology 2004 (Vol. 59A, No. 4, 301-305).

[35] Bowles, J. Shattered: Medawar's Test Tubes and their Enduring Legacy of Chaos, Quarterly Review of Biology 73:3-49. (2000) Presents extensive criticism of Medawar's 1952 paper that is the basis of most traditional theories of aging

[36] Dawkins R. The Selfish Gene, 1976 revised edition 1986. Oxford University Press ISBN: 0-19-286092-5

[37] Goldsmith, T. Aging as an Evolved Characteristic – Weismann's Theory Reconsidered. Medical Hypotheses 2004 62-2 304:308

[38] Skulachev V. Aging is a Specific Biological Function Rather than the Result of a Disorder in Complex Living Systems: Biochemical Evidence in Support of Weismann's Hypothesis. Moscow State University

[39] Reference to be supplied

[40] Gavrilov, L. et al, Evolutionary Theories of Aging and Longevity, The Scientific World JOURNAL (2002) 2, 339–356 ISSN 1537-744X; DOI 10.1100/tsw.2002.96

[41] Bartke, Antebi, The endocrine regulation of aging by insulin-like signals, Science 2003 Feb 28;299(5611):1346-51

[42] Spindler S. Rapid and reversible induction of the longevity, anticancer and genomic effects of caloric restriction. Mech Ageing Dev. 2005 Sep;126(9):960-6. Review. PMID: 15927235

[43] P Eriksson M, Brown W, Gordon L, Glynn M, Singer J, Scott L, Erdos M. Recurrent de novo point mutations in lamin A cause Hutchinson–Gilford progeria syndrome. Nature, May 2003

[44] Gray, Md; Shen, Jc; Kamath-Loeb, As; Blank, A; Sopher, Bl; Martin, Gm; Oshima, J; Loeb, La (Sep 1997). The Werner syndrome protein is a DNA helicase. Nature genetics 17 (1): 100–3. doi:10.1038/ng0997-100. PMID 9288107

[45] Skulachev V. A biochemical approach to the problem of aging: "Megaproject" on penetrating ions, the first results and prospects. Biochemistry (Moscow). 2007. (The SKQ Megaproject)

[46] Valenzano, D, et al. Resveratrol Prolongs Lifespan and Retards the Onset of Age-Related Markers in a Short-Lived Vertebrate. Current Biology 16 296-300 Feb 7 2006.

[47] Goldsmith T. Arguments against non-programmed aging theories. Biochemistry (Moscow) 78:9 971-978 2013

[48] Goldsmith T. The case for programmed mammal aging. Russian Journal of General Chemistry July 2010, Volume 80, Issue 7, pp 1434-1446 DOI: 10.1134/S107036321007039X

[49] Goldsmith T. Rationale for Complex Programmed Lifespan Regulation in Mammals. Homo Sapiens Liberatus Workshop May 2010 Moscow

[50] Kirkwood T, Melov S. On the programmed/ non-programmed nature of ageing within the life history. Current Biology 21, R701–R707, September 27, 2011 DOI 10.1016/j.cub.2011.07.020.

[51] Goldsmith T. On the programmed/ non-programmed aging controversy. Biochemistry (Moscow) Phenoptosis, Vol 77 No 7, pp. 729_732, 2012 DOI: 10.1134/S000629791207005X

[52] Skulachev V. Aging as a particular case of phenoptosis, the programmed death of an organism. (A response to Kirkwood-Melov "On the programmed/ non-programmed nature of aging within the life history"). Aging (Albany NY) 2011

[53] Goldsmith T. Aging by Design. ISBN 978-09788709-3-5 2012 Azinet Press Annapolis

14540052R00109

Printed in Poland
by Amazon Fulfillment
Poland Sp. z o.o., Wrocław